IET MATERIALS, CIRCUITS AND DEVICES SERIES 78

# Energy Harvesting for Wireless Sensing and Flexible Electronics through Hybrid Technologies

**Other volumes in this series:**

Volume 2 **Analogue IC Design: The current-mode approach** C. Toumazou, F.J. Lidgey and D.G. Haigh (Editors)
Volume 3 **Analogue–Digital ASICs: Circuit techniques, design tools and applications** R.S. Soin, F. Maloberti and J. France (Editors)
Volume 4 **Algorithmic and Knowledge-based CAD for VLSI** G.E. Taylor and G. Russell (Editors)
Volume 5 **Switched Currents: An analogue technique for digital technology** C. Toumazou, J.B.C. Hughes and N.C. Battersby (Editors)
Volume 6 **High-frequency Circuit Engineering** F. Nibler *et al.*
Volume 8 **Low-power High-frequency Microelectronics: A unified approach** G. Machado (Editor)
Volume 9 **VLSI Testing: Digital and mixed analogue/digital techniques** S.L. Hurst
Volume 10 **Distributed Feedback Semiconductor Lasers** J.E. Carroll, J.E.A. Whiteaway and R.G.S. Plumb
Volume 11 **Selected Topics in Advanced Solid State and Fibre Optic Sensors** S.M. Vaezi-Nejad (Editor)
Volume 12 **Strained Silicon Heterostructures: Materials and devices** C.K. Maiti, N.B. Chakrabarti and S.K. Ray
Volume 13 **RFIC and MMIC Design and Technology** I.D. Robertson and S. Lucyzyn (Editors)
Volume 14 **Design of High Frequency Integrated Analogue Filters** Y. Sun (Editor)
Volume 15 **Foundations of Digital Signal Processing: Theory, algorithms and hardware design** P. Gaydecki
Volume 16 **Wireless Communications Circuits and Systems** Y. Sun (Editor)
Volume 17 **The Switching Function: Analysis of power electronic circuits** C. Marouchos
Volume 18 **System on Chip: Next generation electronics** B. Al-Hashimi (Editor)
Volume 19 **Test and Diagnosis of Analogue, Mixed-signal and RF Integrated Circuits: The system on chip approach** Y. Sun (Editor)
Volume 20 **Low Power and Low Voltage Circuit Design with the FGMOS Transistor** E. Rodriguez-Villegas
Volume 21 **Technology Computer Aided Design for Si, SiGe and GaAs Integrated Circuits** C.K. Maiti and G.A. Armstrong
Volume 22 **Nanotechnologies** M. Wautelet *et al.*
Volume 23 **Understandable Electric Circuits** M. Wang
Volume 24 **Fundamentals of Electromagnetic Levitation: Engineering sustainability through efficiency** A.J. Sangster
Volume 25 **Optical MEMS for Chemical Analysis and Biomedicine** H. Jiang (Editor)
Volume 26 **High Speed Data Converters** A.M.A. Ali
Volume 27 **Nano-Scaled Semiconductor Devices** E.A. Gutiérrez-D (Editor)
Volume 28 **Security and Privacy for Big Data, Cloud Computing and Applications** L. Wang, W. Ren, K.R. Choo and F. Xhafa (Editors)
Volume 29 **Nano-CMOS and Post-CMOS Electronics: Devices and modelling** Saraju P. Mohanty and Ashok Srivastava
Volume 30 **Nano-CMOS and Post-CMOS Electronics: Circuits and design** Saraju P. Mohanty and Ashok Srivastava
Volume 32 **Oscillator Circuits: Frontiers in design, analysis and applications** Y. Nishio (Editor)
Volume 33 **High Frequency MOSFET Gate Drivers** Z. Zhang and Y. Liu
Volume 34 **RF and Microwave Module Level Design and Integration** M. Almalkawi
Volume 35 **Design of Terahertz CMOS Integrated Circuits for High-Speed Wireless Communication** M. Fujishima and S. Amakawa
Volume 38 **System Design with Memristor Technologies** L. Guckert and E.E. Swartzlander Jr.
Volume 39 **Functionality-Enhanced Devices: An alternative to Moore's Law** P.-E. Gaillardon (Editor)
Volume 40 **Digitally Enhanced Mixed Signal Systems** C. Jabbour, P. Desgreys and D. Dallett (Editors)

Volume 43 **Negative Group Delay Devices: From concepts to applications** B. Ravelo (Editor)
Volume 45 **Characterisation and Control of Defects in Semiconductors** F. Tuomisto (Editor)
Volume 47 **Understandable Electric Circuits: Key concepts, 2nd Edition** M. Wang
Volume 48 **Gyrators, Simulated Inductors and Related Immittances: Realizations and applications** R. Senani, D.R. Bhaskar, V.K. Singh, A.K. Singh
Volume 49 **Advanced Technologies for Next Generation integrated Circuits** A. Srivastava and S. Mohanty (Editors)
Volume 51 **Modelling Methodologies in Analogue Integrated Circuit Design** G. Dundar and M.B. Yelten (Editors)
Volume 53 **VLSI Architectures for Future Video Coding.** M. Martina (Editor)
Volume 54 **Advances in High-Power Fiber and Diode Laser Engineering** Ivan Divliansky (Editor)
Volume 55 **Hardware Architectures for Deep Learning** M. Daneshtalab and M. Modarressi
Volume 57 **Cross-Layer Reliability of Computing Systems** Giorgio Di Natale, Alberto Bosio, Ramon Canal, Stefano Di Carlo and Dimitris Gizopoulos (Editors)
Volume 58 **Magnetorheological Materials and their Applications** S. Choi and W. Li (Editors)
Volume 59 **Analysis and Design of CMOS Clocking Circuits for Low Phase Noise** W. Bae and D.K. Jeong
Volume 60 **IP Core Protection and Hardware-Assisted Security for Consumer Electronics** A. Sengupta and S. Mohanty
Volume 63 **Emerging CMOS Capacitive Sensors for Biomedical Applications: A multidisciplinary approach** Ebrahim Ghafar-Zadeh and Saghi Forouhi
Volume 64 **Phase-Locked Frequency Generation and Clocking: Architectures and circuits for modem wireless and wireline systems** W. Rhee (Editor)
Volume 65 **MEMS Resonator Filters** Rajendra M. Patrikar (Editor)
Volume 66 **Frontiers in Hardware Security and Trust: Theory, design and practice** C.H. Chang and Y. Cao (Editors)
Volume 67 **Frontiers in Securing IP Cores: Forensic detective control and obfuscation techniques** A. Sengupta
Volume 68 **High Quality Liquid Crystal Displays and Smart Devices: Vol. 1 and Vol. 2** S. Ishihara, S. Kobayashi and Y. Ukai (Editors)
Volume 69 **Fibre Bragg Gratings in Harsh and Space Environments: Principles and applications** B. Aïssa, E.I. Haddad, R.V. Kruzelecky and W.R. Jamroz
Volume 70 **Self-Healing Materials: From fundamental concepts to advanced space and electronics applications, 2nd Edition** B. Aïssa, E.I. Haddad, R.V. Kruzelecky and W.R. Jamroz
Volume 71 **Radio Frequency and Microwave Power Amplifiers: Vol. 1 and Vol. 2** A. Grebennikov (Editor)
Volume 72 **Tensorial Analysis of Networks (TAN) Modelling for PCB Signal Integrity and EMC Analysis** Blaise Ravelo and Zhifei Xu (Editors)
Volume 73 **VLSI and Post-CMOS Electronics Volume 1: VLSI and Post-CMOS Electronics and Volume 2: Materials, devices and interconnects** R. Dhiman and R. Chandel (Editors)
Volume 75 **Understandable Electronic Devices: Key concepts and circuit design** M. Wang
Volume 76 **Secured Hardware Accelerators for DSP and Image Processing Applications** Anirban Sengupta
Volume 77 **Integrated Optics Volume 1: Modeling, material platforms and fabrication techniques and Volume 2: Characterization, devices, and applications** G. Righini and M. Ferrari (Editors)

# Energy Harvesting for Wireless Sensing and Flexible Electronics through Hybrid Technologies

Muhammad Iqbal, Brahim Aïssa and Malik Muhammad Nauman

The Institution of Engineering and Technology

Published by The Institution of Engineering and Technology, London, United Kingdom

The Institution of Engineering and Technology is registered as a Charity in England & Wales (no. 211014) and Scotland (no. SC038698).



First published 2023



The Institution of Engineering and Technology
Futures Place
Kings Way, Stevenage
Hertfordshire SG1 2UA, United Kingdom

www.theiet.org



**British Library Cataloguing in Publication Data**
A catalogue record for this product is available from the British Library

**ISBN 978-1-83953-497-3 (hardback)**
**ISBN 978-1-83953-498-0 (PDF)**

Typeset in India by MPS Limited

Cover Image: Guido Mieth/Stone via Getty Images

# Contents

# List of figures

# List of tables

# List of abbreviations

| | |
|---|---|
| AC | Alternating current |
| Au | Gold |
| AlN | Aluminum nitride |
| ADP | Ammonium dihydrogen phosphate |
| Cu | Copper |
| DC | Direct current |
| DRIE | Deep reactive ion etching |
| EDM | Electric discharge machining |
| EM | Electromagnetic |
| EMEH | Electromagnetic energy harvester |
| EMIEH | Electromagnetic insole energy harvester |
| EMF | Electromotive force |
| ESEH | Electrostatic energy harvester |
| FFS | Forward frequency sweep |
| FUC | Frequency-up-conversion |
| GPS | Global positioning system |
| HEH | Hybrid energy harvester |
| HBEH | Hybrid bridge energy harvester |
| HIEH | Hybrid insole energy harvester |
| IoT | Internet of things |
| LED | Light-emitting diode |
| MEMS | Microelectromechanical system |
| NI | National Instruments |
| PE | Piezoelectric |
| PEM-IEH | Hybrid piezoelectric-electromagnetic insole energy harvester |
| PDMS | Polydimethylsiloxane |
| PVDF | Polyvinylidene fluoride |
| PEEH | Piezoelectric energy harvester |
| PZT | Lead zirconate titanate |
| RIE | Reactive ion etching |
| RFS | Reverse frequency sweep |

| | |
|---|---|
| RMS | Root mean square |
| RPM | Revolutions per minute |
| $SiO_2$ | Silicon dioxide |
| SOI | Si-on-insulator |
| TEEHs | Triboelectric energy harvesters |
| WSN | Wireless sensor node |
| WASN | Wireless acceleration sensor node |
| VEHs | Vibration-based energy harvesters |
| WASNs | Wireless acceleration sensor nodes |
| WSNs | Wireless sensors nodes |
| ZnO | Zinc oxide |

# List of symbols

| | |
|---|---|
| $A$ | base acceleration of the vibration |
| $B$ | magnetic flux density |
| $\beta$ | damping coefficient |
| $\xi$ | damping ratio |
| $c$ | damping coefficient |
| $c_{\mathrm{m}}$ | mechanical damping |
| $c_{\mathrm{e}}$ | electrical damping |
| $c_{\mathrm{T}}$ | total damping |
| $d$ | strain constant |
| $d$ | piezoelectric charge constant |
| $d_{\mathrm{n}}$ | piezoelectric strain coefficient |
| $D_{\mathrm{e}}$ | electric displacement |
| $E$ | electric field strength |
| $E$ | Young's modulus of elasticity |
| $\varepsilon$ | dielectric permittivity |
| $f_{\mathrm{n}}$ | natural frequency |
| $F_{\mathrm{r}}$ | restoring force |
| $h$ | thickness of the beam |
| $m$ | magnetic mass |
| $M$ | total mass of the beam and magnet |
| $m_{\mathrm{eq}}$ | equivalent mass |
| $\dot{m}$ | mass flow rate |
| $m_{\mathrm{L}}$ | mass of the beam |
| $m_{\mathrm{T}}$ | tip mass |
| $a$ | area of flow |
| $D_1$ | gap between the coil and magnet |
| $D$ | mean spring diameter |
| $d_{\mathrm{w}}$ | coil wire diameter |
| $G$ | shear stress |
| $H_{\mathrm{m}}$ | height of the magnet |
| $k$ | stiffness matrix |

| | |
|---|---|
| $k_z$ | spring force |
| $N$ | number of coil turns |
| $n$ | number of turns in the spring |
| NdFeB | neodymium |
| $Q$ | quality factor |
| $P$ | power output |
| $P_L$ | load power |
| $R_L$ | load resistance |
| $R_c$ | coil resistance |
| $r_p$ | internal radius of the wound coil |
| $r_m$ | radius of the magnet |
| $r_m$ | radius of magnet |
| $r_p$ | coil inner radius |
| $s$ | mechanical compliance |
| $S$ | strain |
| $S$ | area of coil turns |
| Si | individual coil turn area |
| $t$ | time |
| $T$ | mechanical stress |
| $\omega$ | frequency |
| $\omega_n$ | resonant frequency |
| $\Omega$ | excitation frequency |
| $\rho$ | density of the tip mass |
| $\lambda$ | transmission coefficient |
| $\alpha$ | electromechanical coupling coefficient |
| $x$ | displacement of the mass |
| $\dot{x}$ | velocity of the mass |
| $\ddot{x}$ | acceleration of the mass |
| $y$ | base displacement |
| $\dot{y}$ | velocity of the base |
| $\ddot{y}$ | acceleration of base |
| $z(t)$ | relative displacement between the coil and magnet |
| $U$ | amplitude of velocity |
| $Z$ | displacement of the magnet |

# About the authors

**Muhammad Iqbal** is an assistant professor at the Institute of Computer and Software Engineering, Khwaja Fareed University of Engineering and Information Technology, Rahim Yar Khan, Pakistan. He received his PhD degree in systems engineering with distinction from the Faculty of Integrated Technologies, Universiti Brunei Darussalam, Brunei, in 2020. He holds a Master of Science in mechatronics engineering automation and control from University of Engineering and Technology, Peshawar, gained in 2016, and Bachelor of Science degree in computer engineering from COMSATS University Abbottabad, Pakistan, in 2012. His area of research includes vibration/wind-based energy harvesting, autonomous wireless sensors, MEMS/NEMS devices, biomechanics, and additive manufacturing, among others. He is a professional member of Pakistan Engineering Council, Member IEEE, and approved PhD supervisor by the Higher Education Commission (HEC), Pakistan.

**Brahim Aïssa** earned his PhD in materials science and energy from INRS-EMT, Canada, before joining MPB Communications Inc. where he worked in the Space & Photonics department until 2014. In September 2014, he joined the Qatar Environment and Energy Research Institute (QEERI) as a senior scientist. He collaborated with the EPFL, Switzerland, where he twice hit the world record for power conversion efficiencies in quasi-mono silicon solar cells. He has published more than 200 refereed papers, five books, and holds many patents. Dr. Aïssa has numerous international awards, prizes, and fellowships including the prestigious NPI award from the European Space Agency, the Australian Endeavour fellowship, and the Canadian NSERC R&D industrial fellowship. His PhD and MSc theses were awarded the Top prize for the best scientific achievement, and a part of his work was selected among the best TOP-10 scientific discoveries in Canada. Dr. Brahim is an associate editor in 8 different journals, and a senior fellow of three UNESCO chairs, including the Nanoscience and nanotechnology (UNISA/South Africa), the African Network in Nanoscience (NANOAFNET/South Africa), and the Materials and Technologies for Energy Conversion, Saving and Storage (MATECSS/Canada).

**Malik Muhammad Nauman** is an associate professor at the Faculty of Integrated Technologies, Universiti Brunei Darussalam and has a PhD in Mechatronics Engineering from Jeju National University, South Korea, in 2012. He has published extensively in multidisciplinary areas of 2D/3D printing of energy devices involving functional and smart materials, kinetic energy harvesting via piezoelectric, electromagnetic and hybrid methods, shape memory alloys, and inverse heat and mass transfer. He is a seasoned academician having more than 8 years of teaching undergraduate and graduate courses related to manufacturing, energy engineering, and innovation and entrepreneurship, which enables him to work very closely with students making them skilled engineers, effective innovators, and future entrepreneurs. He is also the ABET Accreditation Coordinator at the Faculty of Integrated Technologies. He is a life-time member of the Pakistan Engineering Council and The American Society of Mechanical Engineers.

# Preface

Harvesting biomechanical energy is a viable solution to sustainable power to wearable electronics for continuous medical health monitoring, remote sensing, and motion tracking. The current research in wearable electronics is trending towards miniaturization, portability, integration, and sustainability. As wearable microelectronics are becoming more ubiquitous, research interest is increasing for the replacement of batteries to harness power from the user's environment by embedded systems. Efforts have been made to prolong the harvester's operational life, overcoming energy dissipation, lowering down of the resonant frequency, attaining multiresonant states and widening the operating frequency bandwidth of the biomechanical energy harvesters. Human body mechanics, especially walking, running, and jogging, can overcome the limitation of nonstop power supply energy to wearable electronics which is, in fact, a constraint for the lasting operation of these devices in well-being monitoring applications.

This book reports on miniature electromagnetic and hybrid piezoelectric-electromagnetic energy harvesters for incorporation into the sole of a commercial shoe to harvest walking into useful electrical energy for the sustainable operation of wearable microelectronics. This book makes contributions pertaining to the design, modeling, fabrication, and characterization of the insole energy harvesters under low-frequency human motions.

An electromagnetic insole energy harvester (EMIEH), capable of efficiently harvesting low-frequency biomechanical energy, has been reported. The harvester shows higher sensitivity to low-frequency external vibrations than conventional cantilever-based design, and hence allows low-impact energy harvesting such as harvesting energy from walking, running, and jogging. The experimentally tested four resonant frequencies appeared at 8.9, 28, 50, and 51 Hz. The harvester has been integrated into the shoe, and a 100 $\mu$F capacitor was charged up to 1 V for about 8 min foot movement.

A multimodal nonlinear hybrid piezoelectric-electromagnetic insole energy harvester (PEM-IEH), converting walking energy into electricity by supplemental conversion mechanism, being able to power wireless sensors used in medical health monitoring for human well-being is reported. The fabricated prototype is quite compact with a volume of 46.8 $cm^3$ and a lightweight of 43.3 g. The response to low-frequency excitations, nonlinear behavior, multiresonant states, and combined piezoelectric-electromagnetic conversion are the key factors that enhance the device performance in capturing more energy. The hybridized harvester can be installed inside a commercial shoe to harvest low-frequency walking energy into electricity for use to power up small electronic devices, such as a health monitoring sensor or a fitness tracker.

A hybrid insole energy harvester (HIEH), capable of harvesting energy from low-frequency walking step motion, to supply power to wearable sensors, has been reported. The multimodal and multi-degree-of-freedom low-frequency walking energy harvester has a lightweight of 33.2 g and occupies a small volume of 44.1 $cm^3$. Experimentally, the HIEH exhibits six resonant frequencies corresponding to the resonances of the intermediate square spiral planar spring at 9.7, 41, 50, and 55 Hz, the polyvinylidene fluoride (PVDF) beam-I at 16.5 Hz and PVDF beam-II at 25 Hz. As compared to individual harvesting units, the hybrid harvester performed much better, generated 7.01 V open-circuit voltage and charged a 100 $\mu$F capacitor up to 2.9 V by hand movement for about 8 min, which is 30% more voltage than the standalone piezoelectric unit in the same amount of time.

This book explores the practical applications of energy harvesting in remote sensing and flexible electronics. It covers a wide range of applications, from environmental monitoring to structural health monitoring and human–machine interfaces. The book also discusses the challenges and opportunities in the commercialization of energy-harvesting technology and their use in space applications.

In summary, this book provides a comprehensive overview of the field of energy harvesting for remote sensing and flexible electronics through hybrid technologies. It covers the basic principles of energy harvesting, the different types of energy sources and devices, power management and conditioning, and the design and optimization of energy-harvesting systems. The book also explores the practical applications of energy harvesting in remote sensing and flexible electronics, and the challenges and opportunities in the commercialization of this technology. It is a valuable resource for researchers, engineers, and students interested in the field of energy harvesting and its application in remote sensing and flexible electronics.

# Acknowledgments

This book would not have been possible without the tremendous support of many people. As authors of *Energy Harvesting for Remote Sensing and Flexible Electronics through Hybrid Technologies*, we would like to express our sincere gratitude to all of the scientists who have contributed to the creation of this project.

First and foremost, we would like to thank our colleagues from the Institute of Computer and Software Engineering in Khwaja Fareed University of Engineering and Information Technology, the Faculty of Integrated Technologies in Universiti Brunei Darussalam, and the Qatar Environment and Energy Research Institute in Hamad Bin Khalifa University, and mentors in the field of hybrid technologies, who have shared their knowledge and expertise with us over the years. We are also grateful for the opportunity to learn from and collaborate with such talented and dedicated individuals.

We would also like to thank the reviewers who generously provided their time and insights to help improve the quality of this book. Their comments and suggestions are invaluable and greatly appreciated.

We wish to underline the valuable help provided by the staff of The Institution of Engineering and Technology (IET) publishing group, with a special mention to Sarah Lynch, Olivia Wilkins, Christoph von Friedeburg, and Paul Deards for their availability and willingness to provide guidance and advice during the preparation of the manuscript. Thank you for having worked closely with us to bring this book to fruition. Your professionalism and dedication made the process of writing and publishing this book a real pleasure.

Finally, we would also like to thank our families and friends, who have supported and encouraged us throughout this process. Your love and encouragement has meant the world to us, and we could not have achieved it without you. It is our hope that this book will be a valuable resource for those interested in the unique challenges and opportunities of deploying energy-harvesting systems through hybrid technologies. We are grateful for the opportunity to share our knowledge and experience with you, and we hope that it will inspire others to pursue their own interests in this exciting and important field of science and technology. Thus, this book is a collaborative effort of many scientists, and we are deeply thankful to everyone who has contributed to its creation.

We are thankful to all people who have created the particular magic environment, where the business objectives are realized through the cultivation of innovation and by harnessing scientific curiosity.

# Chapter 1
# Introduction

With the advent of low-power electronics and improved microfabrication techniques, energy harvesting has received considerable attention over the last decade. The research motivation in this field is the minimal power requirements of electronic devices, such as portable electronic gadgets and wireless sensor nodes (WSNs) used in health-monitoring applications.

In a WSN depicted in Figure 1.1, the sensor detects a physical change in the surrounding environment and sends the signal to a microcontroller for processing. A wireless transceiver transmits data to and receives data from a base station, which is then stored in the memory unit.

The goal of energy harvesting is to power electronic devices by tapping energy sources freely available in the environment and whose output is generally wasted or their potential is not considered adequately. Therefore, by recovering this lost energy through energy harvesting techniques, the repeated battery replacements and maintenance requirements, which are major constraints in the use of many WSNs and portable electronic devices, can be minimized. Vibration-based energy harvesters (VEHs), which convert mechanical vibration energy into electrical energy, have been identified as the type of energy harvesters with the highest potential due to the wide availability of vibration-based energy sources in the environment such as human motion, vibrations in machines, vehicles, buildings, bridges, etc.

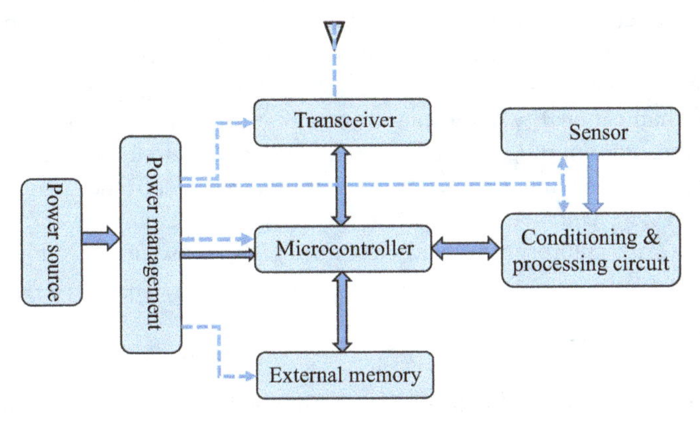


*Figure 1.1 Block diagram of a WSN*

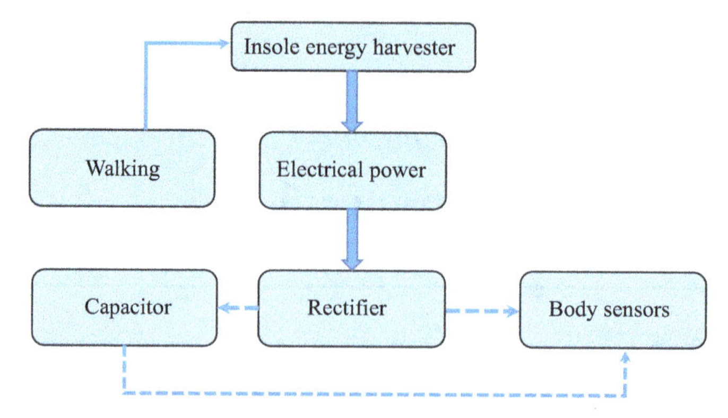


*Figure 1.2 Schematic representation of the adopted energy-harvesting mechanism*

Harvesting biomechanical energy is a viable solution to sustainable power to wearable electronics for continuous medical health monitoring, remote sensing, and motion tracking. The current research in wearable electronics is trending towards miniaturization, portability, integration, and sustainability. As wearable microelectronics are becoming more ubiquitous, research interest is increasing for the replacement of batteries to harness power from the user's environment by embedded systems. Efforts have been made to prolong the harvester's operational life, overcoming energy dissipation, lowering down of resonant frequency, attaining multiresonant states, and widening the operating frequency bandwidth of the biomechanical energy harvesters. Human body mechanics, especially walking, running, and jogging, can overcome the limitation of nonstop power supply energy to wearable electronics which is, in fact, a constraint for the lasting operation of these devices in well-being monitoring applications. This book reports miniature piezoelectric energy harvesters (PEHs), electromagnetic energy harvesters (EMEHs), and hybrid piezoelectric-electromagnetic insole energy harvesters (PEM-IEHs) to harvest walking into useful electrical energy for the sustainable operation of wearable microelectronics. The book makes contributions pertaining to the design, modeling, fabrication, and characterization of the insole energy harvesters under low-frequency human motions.

Biomechanical energy harvesting from walking, running, and jogging provides a considerable amount of kinetic energy that can be converted into electrical energy for sustainably powering microsensors and portable electronics. Biomechanical energy harvesters can be a potential power source and may be integrated with health-monitoring microsensors and electronic gadgets as shown in Figure 1.2.

## 1.1 Background

Kinetic energy harvesting for powering autonomous wireless sensors and microelectronics has been a challenge. Energy harvesting is a process by which ambient energy can be obtained and converted into electricity for small-sized devices such as autonomous WSNs and consumer electronics. Different sources of harvesting energy include energy harvesting from human motion, thermal energy sources, wind, solar, acoustic, and nuclear reaction, etc., but harvesting energy from vibrations is one of the most promising technologies. Among different VEHs, piezoelectric energy harvesters (PEEHs), electromagnetic energy harvesters (EMEHs), and electrostatic energy harvesters (ESEHs) are the most important and efficient, which may provide an affordable, sustainable, and maintenance-free power solution to low-power wireless and portable devices. EM and PE conversion mechanisms are particularly interesting because of the higher electromechanical coupling without the need for initial external voltage sources. VEHs usually generate peak power near the resonant states, thus hindering operation at wide frequency bandwidths. Biomechanical conversion into electrical energy can be accomplished via PE, electrostatic, EM, triboelectric, or hybrid transduction mechanisms. Several methods have been reported in the design of insole energy harvesters to lower down the resonant frequency and broaden the device's frequency bandwidth with the introduction of PE materials. However, due to the miniature design of insole generators, these devices resonate at much higher frequencies $>100$ Hz and need frequency tuning by various techniques, such as frequency-up-conversion. Flexible PE and triboelectric harvesters have been recently reported for harvesting low-frequency body mechanics and textile-based wearable nanogenerators, where moving charges can be induced by polarization and rubbing between an electrode and a dielectric respectively, to bridge the frequency disparity. However, the natural frequencies of most of the previously reported energy harvesters are still on the higher side and inevitably, perform sub-optimally under low-frequency human body vibrations.

## 1.2 Book outline

The main theme of the research reported in this book is representing low-frequency PEEHs, EMEHs, and hybrid PEM-IEHs for application in the sustainable operation of wireless microsensors and microelectronic gadgets for nonstop medical and health monitoring in space applications. The designed models have been simulated, fabricated, and characterized under sinusoidal input vibrations from a novel vibration shaker. The fabricated prototypes were incorporated into the sole of a commercial shoe to evaluate their performance experimentally by operating a pedometer and charging a capacitor. The book is comprised of eleven chapters and each chapter represents a full manuscript, including an introduction, materials and methods, and results and discussion. An overall conclusion of the book is presented in conclusions and future directions, in Chapter 11. References are presented in the bibliography list.

**Chapter 2**, Vibration-based energy harvesting, presents a comprehensive literature review and technical background of VEH. The chapter provides a comprehensive review of recent developments in VEH technologies with a focus on

PEEHs, EMEHs, and hybrid PEEHs. This chapter also introduces PEEHs with respect to device architecture, conversion mechanism, performance parameters, and implementation. The harvesting devices have been compared in terms of generated power, output voltage, resonant frequencies, internal impedance, and base excitation levels to which PEEHs were subjected.

**Chapter 3**, Piezoelectric, electromagnetic, and hybrid energy harvesters, presents a detailed background information about different PEEHs, EMEHs, and hybrid PEM-EHs. An analytical and graphical comparison of the reported energy harvesters has been presented in this chapter on the basis of the device size, voltage and power-generation capability, resonant modes, power densities, and the base excitation to which these energy harvesters have been exposed.

**Chapter 4**, Design and modeling of vibration energy harvesters, presents the analytical modeling, simulation, and finite element analysis (FEA) for hybrid PEM-IEHs. The governing equation for resonant frequencies and length of the cantilever beam is simulated.

**Chapter 5**, Nonlinear 3D printed electromagnetic vibration energy harvesters, describes the design, simulation, fabrication, and characterization of a nonlinear circular-type spiral spring-based EMEH for incorporation into the sole of a commercial shoe. The introduction of the circular spiral spring holding disk magnets above and below the central platform as tip masses has reduced the resonant frequency of the harvester and attained a walking frequency of <10 Hz. The prototype was successfully integrated with the shoe for harvesting biomechanical energy from walking, jogging, and running. The developed harvester exhibits multiresonant frequencies with a wide operating frequency range of about 41 Hz.

**Chapter 6**, Fabrication and characterization of nonlinear multimodal electromagnetic insole energy harvesters, discusses about the design, modeling, fabrication, and characterization of a circular spiral spring-based hybrid PEM-IEH. The developed prototype in this chapter is an extension of the EMIEH presented in Chapter 5. The introduction of an upper and lower piezoceramic plate carrying wound coils made the harvester a hybrid system for an improved electrical output. The harvester was modeled as a mass–spring–damper system and simulated for Eigenfrequency analysis in COMSOL Multiphysics®. The harvester was demonstrated by incorporating into the heel of a shoe and walking on the treadmill.

**Chapter 7**, Design, modeling, fabrication, and characterization of a hybrid piezo-electromagnetic insole energy harvester, presents a multi-degree-of-freedom hybrid PEM-IEH using an upper and lower PVDF cantilever beam as PE power generators. A central square spiral spring is enclosed in Teflon spacers and two PVDF cantilever beams holding wound coils just in line with the magnetic masses ensure dual transduction. The design, simulation, fabrication, and experimental validation of the hybrid harvester are discussed in this chapter.

**Chapter 8**, Multi-degree-of-freedom hybrid piezo-electromagnetic insole energy harvesters, explains about MDOF hybrid PEM-IEHs. Three degree of freedom systems are reported in this chapter.

**Chapter 9**, An overview of the finite element analysis and its applications in kinetic energy harvesting devices, presents the modal analysis of hybrid energy harvesters using COMSOL Multiphysics® software.

**Chapter 10**, Energy harvesters for biomechanical applications, describes about energy harvesters' applications for wearable electronic gadgets and medical health-monitoring sensors. This chapter states ways to make these microsystems autonomous and self-powered for nonstop health monitoring of vital signs in human beings.

**Chapter 11**, Electromagnetic energy harvesters for space applications, provides information about RF and microwave energy harvesting for space applications to monitor and record the health conditions of space mission personnel.

The book concludes with **Chapter 12**, Conclusions and outlook into the future, by presenting the overall results and performance of the developed EMEHs and hybrid insole energy harvesters (HIEHs). A comparison of the fabricated devices with previously reported energy harvesters in literature is also discussed in this chapter. Future recommendations and the proposed research development in the field of insole energy harvesters are also discussed in this chapter.

*Chapter 2*

# Vibration-based energy harvesting

## 2.1 Introduction

Vibration-based energy harvesting (VEH) is a promising solution, particularly, for the sustainable supply of energy to drive microelectronics. Due to the rapid advancement of low-power micro-electro-mechanical systems (MEMS), Internet of Things (IoT), and WSNs, it has attracted significant interest from researchers and industry, to overcome the limitation on the energy storage device or the need to wire power supply to the low-powered devices for operation [1]. Due to the abundant availability of vibration-based energy sources that can be easily harvested, it has been a key focus area. Vibration-based energy sources available in the environment, in the form of biomechanical excitations [2], sinusoidal and random excitations [3,4], rotating excitations [5,6], fluid vertex-induced excitations [7,8], galloping excitations [9], heartbeat excitations [10], and so on, are potential energy sources for providing mechanical energy that can be harvested into electrical energy. Important structural and geometrical parameters that need to be considered and investigated in the design of VEHs include the resonant frequency, bandwidth, damping, volume, weight, and cost of the harvester [11].

A VEH converts ambient vibration energy into useful electricity for low-power WSNs and consumer electronics. To broaden the operation frequency range through multiresonant states, increase the multi-degree-of-freedom, provide nonlinear characteristics, and implement the hybrid conversion of VEHs, different optimization techniques and design considerations are taken into consideration. Important structural and geometrical parameters that need to be considered and investigated in the design of VEHs include the resonant frequency, bandwidth, damping, volume, weight, and cost of the harvester.

VEHs usually generate peak power near the resonant states, thus hindering operation at wide frequency bandwidths [12]. Mechanical vibration energy has the potential to be harvested using several mechanisms including but not limited to piezoelectric (PE), electromagnetic (EM), and hybrid PE-EM conversions.

## 2.2 VEH mechanisms

VEH mainly includes piezoelectric energy harvesters (PEEHs) [13–16], electromagnetic energy harvesters (EMEHs) [17–20], electrostatic energy harvesters

(ESEHs) [21–24], and triboelectric energy harvesters (TEEHs) [25–28]. Generally, in the VEH, ambient mechanical vibration energy is transformed into amplified periodic motion consisting of the conversion of kinetic into potential energy and vice versa, which is then converted into electrical energy with the implementation of a suitable energy transduction mechanism. In the case of PEEHs, the kinetic ambient energy puts the PE materials under stress/strain, resulting in the generation of potential difference (voltage) through an ordered orientation and alignment of the positive and negative charges across the material.

To extract energy from the ambient vibrations, vibration-based PEEHs [29], EMEHs [30], ESEHs [31], and TEEHs [32–39] have been developed and proven to be very capable of providing affordable, sustainable, and maintenance-free power solution for low-power portable and implantable devices. According to research reports, the power production of small-scale VEHs ranges from 10 $\mu$W to 100 mW, and their integration is limited to the applications of low-power sensors and electronics. However, the power generation of relatively large-scale VEHs ranges from 1 W to 100 kW [40].

EM and PE conversion mechanisms are particularly attractive and have higher potential due to their higher electromechanical coupling capability [41] and simple design without the need for an initial external voltage source [42]. However, PEEHs and EMEHs still have some limitations in practical applications [43]. Most of the reported PEEHs and EMEHs use cantilever beam or spring as their motion drive, which suffers from fatigue and damage over time and consequently, shortens the life of the harvesters [44]. PEEHs are also not ideal for implementation as wearable energy harvesters (EHs), as generally, PEEHs respond at much higher input frequencies. Moreover, excessive energy losses occur in PEEHs due to mechanical deformation and single conversion mechanism. A comparison of vibration-based ESEHs, EMEHs, and PEEHs is listed in Table 2.1. Comparatively,

*Table 2.1 Vibration-based energy harvesters' comparison*

| Parameters | ESEHs | EMEHs | PEEHs |
|---|---|---|---|
| Architecture | Difficult | Difficult | Simple |
| Integration with MEMS technology | Moderate | Difficult | Simple |
| Integration with microelectronics technology | Moderate | Difficult | Simple |
| Fabrication | Difficult | Moderate | Simple |
| Micro and nano scaling | Possible | Difficult | Possible |
| Compatibility with integrated electronic devices | Difficult | Difficult | Simple |
| Initial charging | Necessary | Not required | Not required |
| Pull-in phenomenon | Present | Not present | Not present |
| Voltage generation | High level | Low level | High level |
| Current generation | Low level | High level | Low level |
| Internal impedance | Large | Small | Large |
| Natural frequency | High | Low | High |
| AC–DC voltage conversion | Simple | Difficult | Simple |
| Voltage-switching circuit | Necessary | Not required | Not required |
| Sensitivity to vibrations | High | Low | High |
| Power density | High | Low | High |

due to the need for only one PE layer as the transduction medium, PEEHs are simpler to design in any permissible architecture. Moreover, the deposition of the PE layer by the sol–gel process makes the fabrication of PEEHs more suitable to be integrated with standard microelectronics and MEMS technologies. In contrast to ESEHs and EMEHs, the scaling-down (micro and nanoscale development) approach works well for the fabrication of PEEHs. Like ESEHs, the pull-in phenomenon does not occur in PEEHs and also, these do not require an extra battery for the initial charging.

ESEHs and PEEHs can produce relatively higher voltage outputs than those of their counterpart (EMEHs). Moreover, since the output voltage levels that are produced in EMEHs are on the lower side, and as a result, the AC–DC voltage conversion is challenging and usually, EMEHs require special ultra-low voltage and ultra-low power AC–DC convertors. On the contrary to EMEHs, the internal impedances of ESEHs and PEEHs are relatively large because of which small levels of current output are obtainable from both energy harvesters.

Furthermore, one of the major disadvantages of ESEHs is the need for an extra battery for the uninterrupted initial charging of the conductive plates at the beginning of the operational cycle in these harvesters. Moreover, an electrostatic-energy extorting circuit (voltage-switching circuit) is necessary to timely accumulate the transformed energy from conductive plates of the ESEH.

A standalone VEH, irrespective of its type, is currently unable to produce sufficient power levels, even for low-power devices. Hence, the integration of more than one energy transduction mechanism, in the form of hybrid energy harvesters (HEHs), has recently been actively researched, primarily to increase the power levels as compared to the power levels of stand-alone ESEHs, PEEHs, or EMEHs [45], in addition to ensuring broader frequency bandwidth, multiresonant states, and higher output voltage. This chapter reviews the state-of-the-art VEHs, particularly, PEEHs, EMEHs, and hybrid EHs.

## 2.3 Wireless sensor nodes (WSNs)

WSNs and portable electronic gadgets have recently attracted a great deal of attention and interest from researchers. In the last decade, WSN usage and applications in industry and urban infrastructure have gained tremendous attraction. These WSNs can be installed in remote locations, for instance, as structural monitoring sensors, global positioning systems (GPS) tracking devices on animals in the wild as well as human body sensors. The power consumption of most of the microsensors is in the range of microwatts ($\mu$W), as shown in Table 2.2, for pressure, temperature, and acceleration microsensor types. Acceleration microsensors consume more power, as compared to pressure and temperature microsensors. Wireless acceleration sensor nodes (WASNs) may be used to monitor and record the acceleration levels of vibrating bodies and can be installed in critical infrastructure and high-value machines for continuous monitoring.

Some features of commercially available WASNs are presented in Table 2.3. An acceleration sensor can be single, dual, or triaxial, with the ability to detect base

*Table 2.2 Power requirement for different monitoring microsensors [46]*

| Sensor type | Minimum current ($\mu$A) | Minimum voltage (V) | Power ($\mu$W) |
|---|---|---|---|
| Pressure | 1–4 | 1.8–2.1 | 1.8–8.4[a] |
| Temperature | 0.9–14 | 2.1–2.7 | 1.89–37.8[a] |
| Acceleration | 10–180 | 1.8–2.5 | 21.6–324[a] |

[a]Calculated using the equation, $P = VI$.

acceleration up to ±6 g. The operational voltage of the reported mm-scale sensors varies from 0.2 V to 5.5 V, requiring power in the range of 2.2 mW–500 mW for its operation and data transmission.

## 2.4 Traditional electrochemical batteries as a power source for WSNs

Usually, WSNs are operated with batteries. Different types of batteries and other energy storage media in terms of energy densities, specific energy, operating temperature, workable life period, and nominal voltages are presented and compared in Table 2.4. As can be seen, the nominal voltage levels of most super capacitors and thin-film batteries are above 2.5 and 3.5 V, respectively. Li-ion batteries are increasingly being used due to their wide range in terms of a nominal voltage level (1.5–3.7 V), energy density (435–1 560 Wh/l), and specific energy (211–490 Wh/kg). Despite the advances in battery technology, batteries have a limited lifespan and the need to be regularly replaced or recharged. This somewhat limits the use of battery-operated WSNs in embedded, deserted, hazardous, and remote locations.

## 2.5 Potential alternative sources to batteries

Among the reported ambient energy sources available in the environment, harvesting energy from mechanical vibrations is one of the most promising technologies, due to the abundant presence of vibrations in the environment, from moving bodies such as humans [69], machines [70], vehicles [71], bridges [72], buildings [73], ocean waves [74], and other civil entities. Furthermore, kinetic energy harvesters have better power-producing capability, thus allowing sustainable operation of microsystems. Numerous vibrating structures present in the environment are potential sources of energy. Table 2.5 lists some of these vibrating bodies, with their corresponding vibration accelerations and frequencies. The base accelerations of these entities range from 0.01 to 10 g, with a frequency range from 1 to 20 kHz.

An alternative to replacing batteries of the WSNs, either for direct battery replacement or to facilitate battery recharging, is by looking into energy harvesting to directly provide power. Energy harvesting is a technique by which ambient

*Table 2.3 Commercially available WASNs*

| Trade name | Axis | Acceleration range (g) | Bandwidth (kHz) | Operating voltage (V) | Sensitivity (mV/g) | Operating temperature (°C) | Stand-by current ($\mu$A) | Transmission current (mA) | Power (mW) | Size (mm) | Ref. |
|---|---|---|---|---|---|---|---|---|---|---|---|
| KXPA4 | 3 | ±6 | 0–3 300 | 2.8[b] | — | −40–85 | <10[b] | 1.1 | 35.08[a] | — | [47] |
| 8732A500 | 3 | ±5 | 76 | 4[b] | 20 | −50–120 | 2–18[b] | 2 | 53.48[a] | — | [48] |
| ADXL202 | 2 | 2 | — | 0.2[b] | 167 (3V) 312 (5V) | −40–85 | 0.025[b] | 13.7 | 3–5.25[a] | 5 × 5 × 2 | [49] |
| ADXL335 | 3 | ±3 | 550–1 600 | 1.35–1.65[b] | 270–330 | −40–85 | 0.06[b] | 0.035 | — | 4 × 4 × 1.45 | [50] |
| KYL500S | — | — | <25 | 5[b] | — | −40–85 | <20[b] | <40 | 500[a] | — | [51] |
| JA30SA32-25B | — | 1.5 | — | — | 1 000 | −10–60 | — | — | 4.75–5.25[a] | 14 × 11 ×5 | [52] |
| MXS2200E | | 1 | — | — | 2 000 | −40–85 | — | — | 2.7–5.25[a] | 5 × 5 ×2 | [53] |
| H48C | 3 | ±3 | — | 4.5–5.5[b] | 333 | −25–75 | 7–10[b] | — | 2.2–3.6[a] | 4.8 × 4.8 × 1.5 | [54] |
| Model 1221 | 1 | ±2 | 0–400 | 0–4[b] | 2 000 | −55 to +125 | 8–10[b] | — | 50[a] | 3.5 × 3.5 × 1.05 | [55] |
| LIS3L02AQ | 3 | ±2.2–6 | 2.5–4 | 3–5.25[b] | 660 | −40–85 | — | 1 | 2.4–3.6[a] | 7 × 7 × 1.8 | [56] |

[a]Theoretical value.
[b]Experimentally measured value.

*Table 2.4 Comparison of different storage devices*

| Device | Make | Nominal voltage (V) | Energy density (Wh/l) | Specific energy (Wh/kg) | Operating temperature (°C) | Workable life (h) | Ref. |
|---|---|---|---|---|---|---|---|
| Li-ion batteries | TENERGY | 3–3.70 | 435 | 211 | −20–50 | 2 000 | [57] |
| | MULTICOMP | 1.5 | 1 270–1 560 | 370–490 | –20–45 | 45 000 | [58] |
| | HICHARGE | 3.7 | — | — | 0–45 | 672 | [59] |
| Thin-film batteries | ST LIFE | 3.70 | <50 | <1 | −20–70 | >1 000 | [60] |
| | THINERGY | 3.9 | 4 | 300 | −40–85 | 100 000 | [61] |
| | MICRO-BATTERY | 4.2–4.5 | 1 | 400 | −40–150 | <100 0000 | [62] |
| Super capacitors | MAXWELL | 1.25 | 6 | 1.5 | −40–65 | >10 000 | [63] |
| | DCN | 2.7 | 4.2972 | 2.5859 | −40–60 | 1 000 | [64] |
| | MAXWELL | 2.7 | 0.025 | 3.4 | −40–65 | 1 000 | [65] |
| | NESSCAP | 2.7 | 0.051 | 4.51 | −40–65 | 50 000 | [66] |
| | NaNiCl | 2.6 | 250–270 | 100–200 | 270 | 80–90 | [67] |
| | MAXWELL | 2.70 | 0.66 | 4.1 | −40–65 | 1 500 | [68] |

*Table 2.5 Acceleration and frequencies of different vibrating bodies*

| **Vibration source** | | **Acceleration (g)** | **Frequency (Hz)** | **Ref.** |
|---|---|---|---|---|
| Bridge | | 0.01–3.79 | 1–50 | [75] |
| Human body | Walking | 5–10 | <20 | [76] |
| | Tapping heel | 3 | 1 | [77] |
| Home appliances | Refrigerator | 0.01 | 240 | [78] |
| | Liquidizer casing | 0.65 | 121 | [79] |
| | Car AC system | 0.08–0.17 | 20–20k | [80] |
| | Microwave oven | 0.22 | 121 | [81] |
| | Base of a 3-axis machine | 10 | 70 | [82] |
| | Washing machine | 0.05 | 109 | [83] |
| | Computer | 0.06 | 75 | [84] |
| Aircrafts | | 10 | 1–1 000 | [85] |
| | Symmetric wing bending | 0.011 | 4.92 | |
| | Asymmetric wing bending | 0.032 | 10.47 | |
| Equipment | HVAC vents | 0.02–0.15 | 60 | [86] |
| Environmental bodies | Wooden stairs in walking | 0.13 | 385 | [87] |

*Table 2.6 Comparison of different energy sources*

| **Energy source** | **Power density (mW/cm$^3$)** | **Energy density (mWh/cm$^3$)** | **Ref.** |
|---|---|---|---|
| Nuclear reaction | 80 | $1 \times 10^6$ | [98] |
| Solar | 15 | — | [99] |
| Vibrations | 10–300 | — | [100] |
| Airflow | 0.01–0.1 | — | [101] |
| Batteries (Li-ion) | 90 | 300 | [102] |
| Batteries (zinc) | — | 1 050–1 560 | [103] |
| Acoustic noise (100–160 dB) | 0.00096 | $2.94 \times 10^{-12}$<br>$14.48 \times 10^{-3}$ | [104] |
| Human motion | 0.1–1 | — | [105] |
| Thermoelectric (10% gradient) | 0.040 | — | [106] |

energy can be converted into beneficial electricity, particularly for low-power electronic devices such as WSNs and consumer electronics. Different means of energy sources may be harvested, including energy from human motion [88], rotating tires [89], thermal [90], wind [91,92], solar [93], acoustic [94], nuclear reaction [95], and bioenergy [96,97]. A comparison of different energy sources and their power and energy densities is presented in Table 2.6.

# Chapter 3

# Piezoelectric, electromagnetic, and hybrid energy harvesters

## 3.1 Introduction

Among the reported ambient energy sources available in the environment, harvesting energy from mechanical vibrations is one of the most promising technologies, due to the abundant presence of vibrations in the environment, from moving bodies such as humans [69], machines [70], vehicles [71], bridges [72], buildings [73], ocean waves [74], and other civil entities. Furthermore, KEHs have better power-producing capability, and thus, allowing sustainable operation of microsystems. Numerous vibrating structures present in the environment are potential sources of energy. Table 3.1 lists some of these vibrating bodies, with their corresponding vibration accelerations and frequencies. The base accelerations of these entities range from 0.01 to 10 g, with a frequency range from 1 to 20 kHz.

*Table 3.1 Acceleration and frequencies of different vibrating bodies*

| **Vibration source** | | **Acceleration (g)** | **Frequency (Hz)** | **Ref.** |
|---|---|---|---|---|
| Bridge | | 0.01–3.79 | 1–50 | [75] |
| Human body | Walking | 5–10 | <20 | [76] |
| | Tapping heel | 3 | 1 | [77] |
| Home appliances | Refrigerator | 0.01 | 240 | [78] |
| | Liquidizer casing | 0.65 | 121 | [79] |
| | Car AC system | 0.08–0.17 | 20–20k | [80] |
| | Microwave oven | 0.22 | 121 | [81] |
| | Base of a 3-axis machine | 10 | 70 | [82] |
| | Washing machine | 0.05 | 109 | [83] |
| | Computer | 0.06 | 75 | [84] |
| Aircrafts | | 10 | 1–1 000 | [85] |
| | Symmetric wing bending | 0.011 | 4.92 | |
| | Asymmetric wing bending | 0.032 | 10.47 | |
| Equipment | HVAC vents | 0.02–0.15 | 60 | [86] |
| Environmental bodies | Wooden stairs in walking | 0.13 | 385 | [87] |

To extract energy from the ambient vibrations, vibration-based PEEHs [29], EMEHs [30], ESEHs [24], and TEEHs [32–39] have been developed and proven to be very capable of providing affordable, sustainable, and maintenance-free power solution for low-power portable and implantable devices. According to research reports, the power production of small-scale VEHs ranges from 10 $\mu$W to 100 mW, and their integration is limited to the applications of low-power sensors and electronics. However, the power generation of relatively large-scale VEHs ranges from 1 W to 100 kW [40].

EM and PE conversion mechanisms are particularly attractive and have higher potential due to their higher electromechanical coupling capability [41] and simple design without the need for an initial external voltage source [42]. However, PEEHs and EMEHs still have some limitations in practical applications [43]. Most of the reported PEEHs and EMEHs use cantilever beam or spring as their motion drive, which suffers from fatigue and damage over time and consequently, shortens the life of the harvesters [44]. PEEHs are also not ideal for implementation as wearable energy harvesters (EHs), as generally, PEEHs respond at much higher input frequencies. Moreover, excessive energy losses occur in PEEHs, due to the mechanical deformation and single conversion mechanism. A comparison of vibration-based ESEHs, EMEHs, and PEEHs is listed in Table 3.2. Comparatively, due to the need for only one PE layer as the transduction medium, PEEHs are simpler to design in any permissible architecture. Moreover, the deposition of the PE layer by the sol–gel process makes the fabrication of PEEHs more suitable to be integrated with standard microelectronics and MEMS technologies. In contrast to ESEHs and EMEHs, the scaling-down (micro and nanoscale development)

*Table 3.2 Vibration-based EHs' comparison*

| Parameters | ESEHs | EMEHs | PEEHs |
|---|---|---|---|
| Architecture | Difficult | Difficult | Simple |
| Integration with MEMS technology | Moderate | Difficult | Simple |
| Integration with microelectronics technology | Moderate | Difficult | Simple |
| Fabrication | Difficult | Moderate | Simple |
| Micro and nano scaling | Possible | Difficult | Possible |
| Compatibility with integrated electronic devices | Difficult | Difficult | Simple |
| Initial charging | Necessary | Not required | Not required |
| Pull-in phenomenon | Present | Not present | Not present |
| Voltage generation | High level | Low level | High level |
| Current generation | Low level | High level | Low level |
| Internal impedance | Large | Small | Large |
| Natural frequency | High | Low | High |
| AC–DC voltage conversion | Simple | Difficult | Simple |
| Voltage-switching circuit | Necessary | Not required | Not required |
| Sensitivity to vibrations | High | Low | High |
| Power density | High | Low | High |

approach works well for the fabrication of PEEHs. Like ESEHs, the pull-in phenomenon does not occur in PEEHs and also, these do not require an extra battery for the initial charging.

ESEHs and PEEHs can produce relatively higher voltage outputs than their counterparts (EMEHs). Moreover, since the output voltage levels that are produced in EMEHs are on the lower side, and as a result, the AC–DC voltage conversion is challenging and usually, EMEHs require special ultra-low voltage and ultra-low power AC–DC convertors. On the contrary to EMEHs, the internal impedances of ESEHs and PEEHs are relatively large because of which small levels of current output are obtainable from both EHs.

Furthermore, one of the major disadvantages of ESEHs is the need for an extra battery for the uninterrupted initial charging of the conductive plates at the beginning of the operational cycle in these harvesters. Moreover, an electrostatic-energy extorting circuit (voltage-switching circuit) is necessary to timely accumulate the transformed energy from conductive plates of the ESEH.

Stand-alone VEH, irrespective of its type, is currently unable to produce sufficient power levels, even for low-power devices. Hence, the integration of more than one energy transduction mechanism, in the form of HEHs, has recently been actively researched, primarily to increase the power levels as compared to the power levels of stand-alone ESEHs, PEEHs, or EMEHs [45], in addition to ensuring broader frequency bandwidth, multiresonant states, and higher output voltages. This chapter reviews the state-of-the-art VEHs, particularly, PEEHs, EMEHs, and hybrid EHs. Section 3.2 overviews theories of PE and EM, before indulging into different PEEHs and EMEHs that have been developed in the literature. This section also reviews the different hybrid EHs which combine different harvesting mechanisms. Section 3.3 discusses the different EHs that are available in the literature, so as to compare and analyze the different EHs. Section 3.4 summarizes the chapter by highlighting the main points with future recommendation remarks.

## 3.2 Vibration-based energy harvesting

VEH is a viable solution to sustainable power microelectronics. Recent advancements in WSNs, MEMS, and IoTs, energy harvesting from ambient environment have attracted great research interest as most IoT devices run on batteries that have a limited lifespan, and they need to be replaced or recharged regularly [107]. A solution to this is to use energy-harvesting technologies such as solar, thermal, or mechanical to generate energy for the device. This reduces the need for frequent battery replacements and increases the lifespan of the device. Mechanical vibration energy has the potential to be harvested using several mechanisms including but not limited to PE, EM, and hybrid PE- EM conversion [108].

### *3.2.1 Piezoelectric energy harvesters*

PE materials are commonly used in VEHs, due to the interaction between mechanical and electrical states. The crystalline structure of PE materials allows them to transform mechanical strain energy into electrical charge and to convert an applied electrical potential into the mechanical strain. This conversion property provides its ability to extract mechanical energy from the surroundings, usually ambient vibrations, and transforms it into useful electrical energy that can be utilized to operate WSNs. Vibration-based PEEHs are mostly composed of a cantilever beam with a proof mass attached to the beam. Oscillation of the beam due to ambient vibrations allows the conversion of mechanical energy into electrical energy by the direct PE effect, which is then transferred to an external circuit. The PE effect exists in two reversible domains: direct conversion of mechanical strain to electrical charge in which the material behaves like a sensor, and the reverse, in which mechanical strain is induced by an applied electric field, much like an actuator [109]. Only the mechanical to electrical conversion is directly applicable to the operation of the EH. Of course, part of the mechanical energy is also dissipated in various forms of damping, like air damping, dielectric losses, and material internal damping. PEEHs have a relatively higher energy density than both EMEHs and ESEHs. The simple constitutive equations [110] for PE materials can be formulated as in (3.1) and (3.2).

$$S_m = s_i T + d_n E \tag{3.1}$$

$$D_e = d_j T + \varepsilon_m E \tag{3.2}$$

where $S_m$ is the mechanical strain, $s_i$ is the compliance, $T$ is the mechanical stress, $D_e$ is the electric displacement, $E$ is the electric field, $d_n$ is the PE strain coefficient, and $\varepsilon_m$ is the dielectric permittivity (between the electric displacement and electric field) at zero mechanical stress. The performance of a VEH can significantly be influenced by nonlinearity in PE coupling, and an increase in nonlinear PE coefficient increases the optimal power [111].

#### 3.2.1.1 PE materials

PE materials generate electrical energy when subjected to mechanical strain as a result of the applied force or pressure. The generated energy in PE materials is proportional to the applied mechanical energy. A variety of materials exhibit PE effect, including the naturally occurring crystalline (quartz and zinc oxide), polymers (polyvinylidene fluoride (PVDF)), ceramics and PE composites (lead lanthanum zirconate titanate ($Li_2Si_2O_5$), barium titanate ($Ba_2TiSiO_6$), lead zirconate titanate (PZT)), and artificially developed crystals (barium titanate, ammonium dihydrogen phosphate, and gallium orthophosphate). The most commonly used PE materials are polymers and ceramics, which are presented in Table 3.3.

PZT offers several advantages in MEMS applications because of its compatibility with microfabrication techniques and the ease with which it is fabricated with the sol–gel process. Moreover, it provides a good signal-to-noise ratio in a wide dynamic range and due to its high energy density, the PE coefficient and dielectric constant values are superior to ZnO and AlN. Polymer and ceramic PE materials are widely used in PEEHs; the latter is capable of producing relatively more power than the former because of better PE properties and high dielectric constant values [112].

*Table 3.3 Properties of common PE materials*

| PE material | Symbol | Commercial name | PE constant (pm/V) | Crystalline form |
|---|---|---|---|---|
| Lead zirconate titanate | $Pb\{Zr, Ti\}O_3$ | PZT | 117[b] | Polycrystalline ceramic |
| Polyvinylidene fluoride | $-(C_2H_2F_2)_n-$ | PVDF | 28[c] | Oriented film |
| Gallium orthophosphate | $GaPO_4$ | | 4.5[a] | Single crystal |
| Lead lanthanum zirconate titanate | $Pb_{1-x}La_x\{Zr_{0.54}Ti_{0.46}\}O_3$ | PZT | 545[b] | Polycrystalline ceramic |
| Ammonium dihydrogen phosphate | $H_6NO_4P$ | ADP | 48[d] | Single crystal |
| Barium titanate | $BaTiO_3$ | | 587[e] | Single crystal |
| Quartz | $SiO_2$ | | 2.3[a] | Single crystal |
| Zinc oxide | ZnO | | 12[b] | Single crystal |

[a]$d_{11}$, [b]$d_{33}$, [c]$d_{31}$, [d]$d_{36}$, and [e]$d_{15}$.

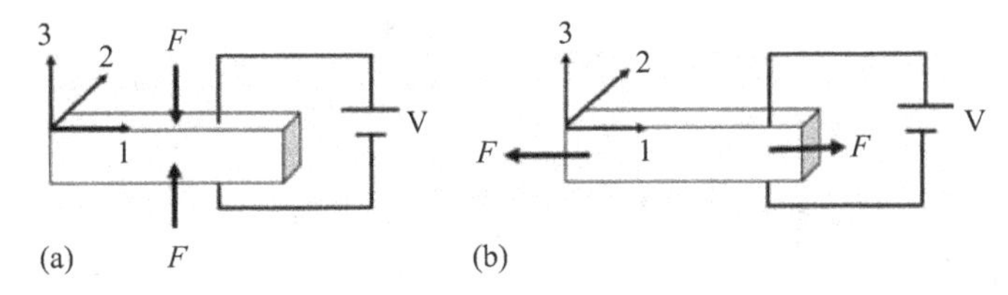


*Figure 3.1 PE conversion modes: (a)* $d_{33}$ *mode and (b)* $d_{31}$ *mode*

Figure 3.1(a) and (b) depicts the conversion modes $d_{33}$ and $d_{31}$, respectively. In the conversion mode $d_{33}$, polarization is induced in direction 3 from the application of mechanical stress applied, also acting in direction 3. Polarization is also induced in direction 3 in conversion mode $d_{31}$; however, this is a result of the application of mechanical stress applied in direction 1, i.e., perpendicular to the induced polarization. For conversion mode $d_{15}$, polarization is induced in direction 1 as a result of the shear stress applied in direction 2 (subscripts 4–6 correspond to the shear stress about directions 1–3, respectively).

From the above, the conversion modes $d_{33}$ and $d_{31}$ generate electrical energy from horizontal and vertical compressive forces, respectively, while the conversion mode $d_{15}$ generates electrical energy from the shear stress effect. Under identical testing conditions, the $d_{33}$ mode is capable of producing more energy as compared to the $d_{31}$ mode, due to its higher PE coefficient and electromechanical coupling [113]. The electrical energy can be increased either by using the coupling mode more efficiently or by increasing the applied stress or strain.

#### 3.2.1.2 Device architecture of the PEEH

In the PEEH, the oscillating cantilever beam or the membrane holding the PE material produces an AC voltage due to alternating polarization in the material when subjected to changing mechanical strain or deformation. This induced AC voltage needs to be rectified into DC voltage and carried to the load or to an ultra-low power sensor. The selection of the PE material depends on several factors including the material elasticity, temperature, and the dielectric constant. The commonly adopted PEEH architectures are unimorph, bimorph, membrane-type, or multibeam cantilever type, as shown in Figure 3.2. An unimorph cantilever type usually carries the PE material on its top side, while a bimorph cantilever type holds PE materials on both sides (top and bottom of the cantilever) to ensure a better output. In a membrane-type EH, the PE layers can be pasted either on top or bottom of the membrane. Commonly, the copper (Cu) or gold (Au) layer is used as an electrode on the PE material for electric transmission.

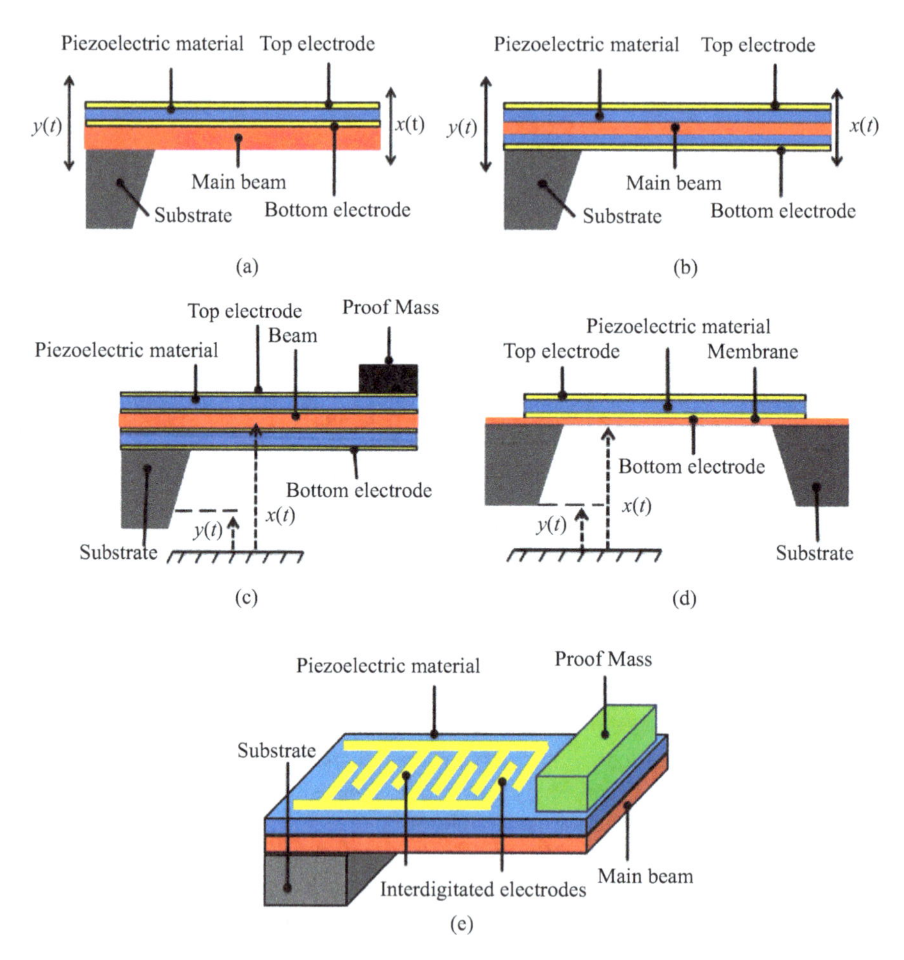


*Figure 3.2 Architectures of resonant PEEHs: (a) unimorph with no proof mass, (b) bimorph with no proof mass, (c) bimorph with proof mass, (d) membrane type, and (e) beam type with interdigitated electrodes*

To obtain optimum output from the PEEHs, the designed device needs to be classified into either resonant or nonresonant PEEH, according to their characteristics. Resonant PEEHs are further classified into linear and nonlinear harvesters [114]. Resonant PEEHs respond at larger amplitudes when the excitation frequency matches the harvester's natural frequency [115]. On the other hand, nonresonant PEEHs are specifically designed to operate in a narrowband or broadband frequency vibration environment and to function relatively over a wide frequency range. Most of the harvesters reported in the literature are resonant type; however, a few nonresonant type harvesters have also been developed.

Linear resonant PEEHs operate in the narrowband frequency range and need tuning for efficient energy extraction from an available excitation frequency range [116]. These energy harvesters exhibit resonance in a small frequency band and are ill suited for off-resonant operations. Generally, linear resonant PEEHs are superior to nonlinear EHs only when subjected to harmonic excitations at their resonant frequencies [117]. However, nonlinear EHs are considered the most efficient [118], as they operate over a wider range of available frequencies and are capable of delivering more output power in practical applications [119]. These harvesters are capable of responding quickly and efficiently to the changes in the frequency of vibrations at larger amplitudes, extracting more power without tuning due to their broad frequency bandwidth [120].

A PE-type microscale harvester, fabricated in [121], is shown in Figure 3.3. During fabrication, AlN and Al layers were deposited on $SiO_2$ by the deep reactive ion etching (DRIE) process, with the moving structure defined from the bottom and front. The $SiO_2$ layer was then removed by selective etching. The 2 $mm^3$ harvesters were fabricated mainly from AlN and PZT with a 1.57 $\mu$g Si tip mass attached to the cantilever beam. The AlN layer on a $SiO_2$ substrate was patterned to define bonding pads, electrical connections, and the top electrode. Finally, the Si layer was removed by the DRIE process. The harvester was mounted on a vibration shaker (as shown in Figure 3.3(b)) and is capable of producing 38 nW power across the matching load resistance at a resonant frequency of 204 Hz under 0.5 g base acceleration.

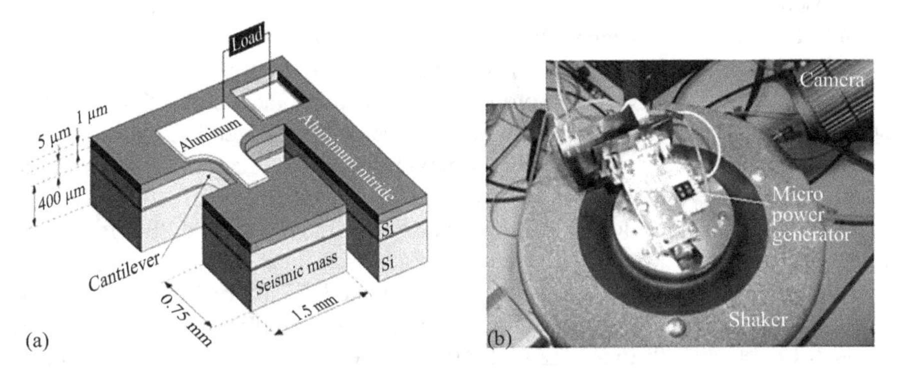


*Figure 3.3 (a) Schematic of the proposed harvester and (b) harvester on the vibration shaker. Reproduced with permission from [121].*

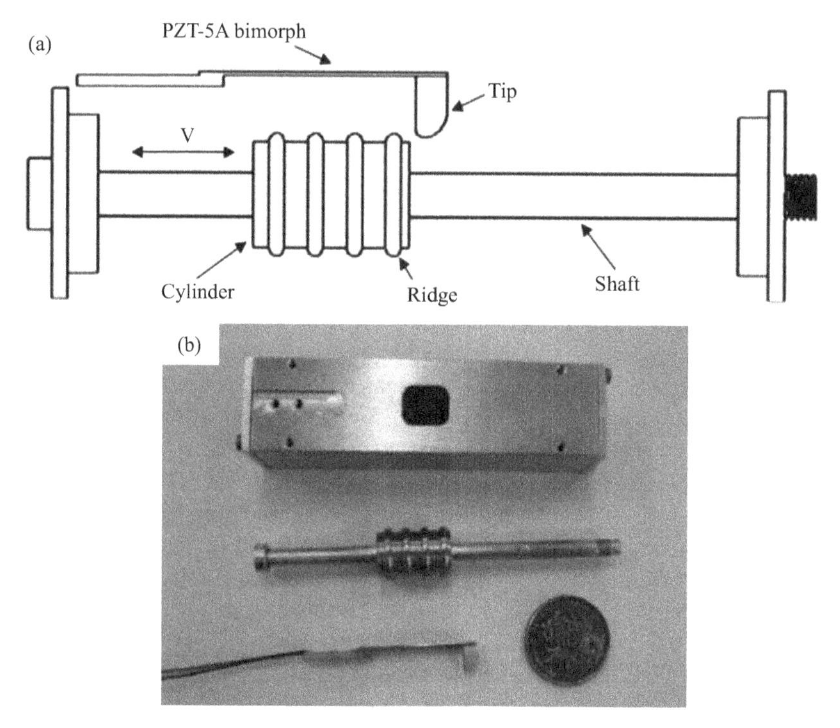


*Figure 3.4 (a) Schematic of the impact-driven PEEH and (b) fabricated prototype. Reproduced with permission from [122].*

A PE-type human motion-driven EH is reported in [122]. The device uses a high-resonant frequency PZT-5A bimorph cantilever beam with a proof mass at the free end. A ridged cylinder enclosed the shaft, and a bimorph with a tip mass was attached to one end of the shaft, as shown in the schematic diagram in Figure 3.4 (a). When the cylinder experiences an impulse excitation, the PZT-5A bimorph starts oscillating due to striking with multiple ridges, which results in an electrical output. The harvester is encapsulated in an aluminum rectangular casting and a rubber pad is used as protection for the fragile PZT at the time of impulse excitation. The size of the developed prototype is 86 400 $mm^3$ and weighs 234 g. With an open-circuit voltage of 2.47 V and when connected across a 20 kΩ optimum load, the harvester is capable of generating a maximum power of 51 $\mu$W from a normal walking speed of 5 km/h on the treadmill, at a resonant frequency of 260 Hz.

Liu *et al.* [123] developed MEMS-based, wide frequency band PEEHs with the capability of converting low-frequency vibrations to high-frequency self-oscillations by the frequency-up-conversion (FUC) method. In the first energy harvester (EH-I) using the FUC technique, the lower resonant cantilever was quenched to broaden the bandwidth to 22 Hz under 0.8 g acceleration, with the high-resonant beam shifted to vibrate at 618 Hz. The second energy harvester (EH-II) is shown in Figure 3.5(b). Its frequency was shifted from 22 to 26 Hz, to achieve resonance,

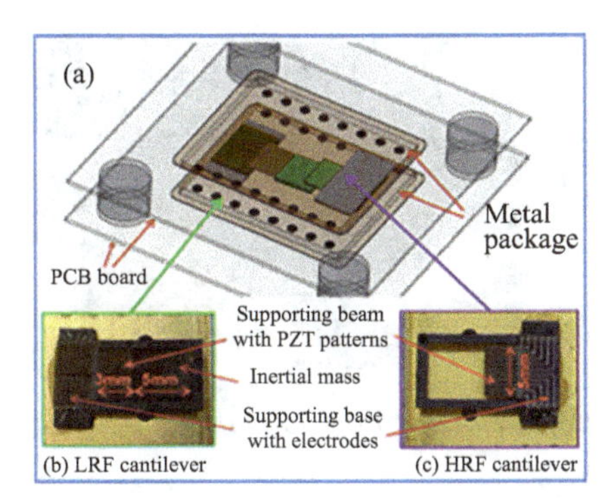


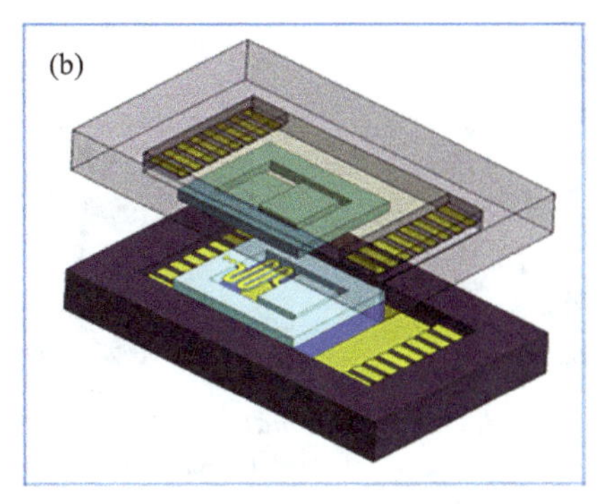


*Figure 3.5 (a) Schematic representation of EH-I reported in [123] and (b) schematic representation of EH-II*

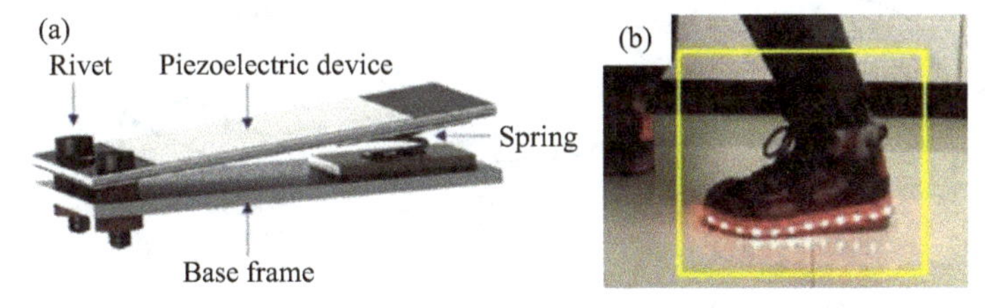


*Figure 3.6 (a) Schematic of the reported PEEH and (b) LED lights switched on during the walk. Reused with permission from [124].*

and using the FUC technique, at lower operating frequencies of 20 and 25 Hz, power densities were enhanced to 61.5 and 159.4 $\mu$W/cm$^3$, respectively, under 0.8 g acceleration level.

A PEEH, which has the potential to be a power source for LED lights installed on commercial shops for road-side night workers, is reported in [124]. The harvester is composed of a piezo-ceramic plate fixed at one end to a base frame, with its free end attached to a spring. Figure 3.6(a) shows the schematic of the PEEH. The PEEH is capable of easily converting biomechanical energy from walking, running, and jumping into electricity. With a lightweight of 14 g and miniature-size (6 mm $\times$ 4 mm $\times$ 3 mm), the harvester can deliver 800 $\mu$W power and 12 V RMS voltage across a matching impedance of 400 k$\Omega$. The harvester is supplied with an LED switching circuit and a provision in the circuit to cut off current when there is no motion.

Isarakorn *et al.* [125] modeled, simulated, and experimentally validated low-resonant MEMS-scale PEEH, based on the epitaxial $Pb(Zr_{0.2}Ti_{0.8})O_3$ PE thin-film unimorph. As shown in Figure 3.7, the PE unimorph was developed on the Si substrate through oxide layers of low dielectric constants and higher PE coefficients. The harvester was simulated for Eigen frequencies and voltage analysis using COMSOL Multiphysics® software. At a resonant frequency of 2.3 kHz, the

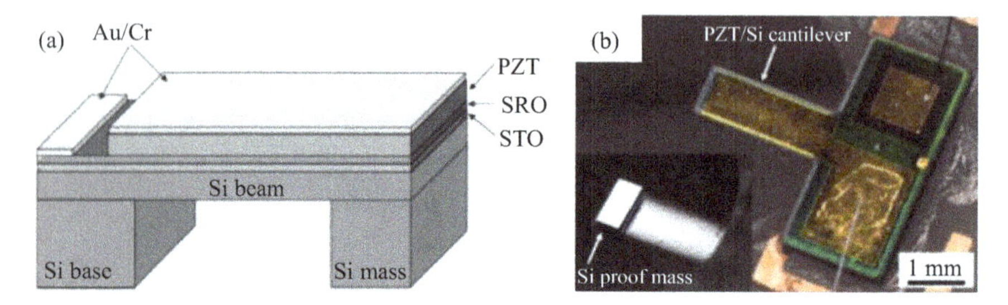


*Figure 3.7 (a) Schematic of the PEEH in [125] and (b) an optical image of the developed harvester*

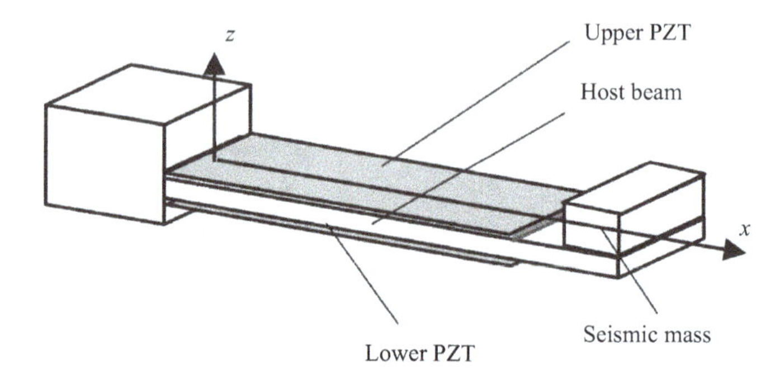


*Figure 3.8 Laminated PEEH reported in [126]*

PEEH produced a current, voltage, and power of 48 $\mu$A, 270 mV, and 13 $\mu$W, respectively, across an optimum load of 5.6 kΩ.

Lu *et al.* [126] presented an analytical model for the conversion efficiency of PEEHs using PZT-PIC255 and PZN-8% PT, as shown in Figure 3.8. The output performance of the harvester using both PE materials has been measured at high frequencies between 2 900 Hz and 7 000 Hz across various load resistances (1–500 kΩ). With an increase in input frequency, an increase in output power has been reported in the case of PZN-8% PT, whereas the output of PZT-PIC255 has shown more variations with a change in the resistive load. The PZT-PIC255 material internal resistance was found to be 68 kΩ and is capable of delivering a peak power of 660 $\mu$W, while PZN-8% PT is capable of generating a maximum power of 571 $\mu$W across 215 kΩ. Comparing the overall efficiencies of both generators, the harvester using PE material PZT-PIC255 has been shown to have better performance than that of the harvester using the PZN-8% PT layer.

A human wearable cantilever-type bimorph PEEH has been reported in [127], for lower-frequency applications. Using the FUC mechanism, the device has been tested for powering up an MP3 player and a mobile phone. Initially, the harvester was modeled analytically with a resonant frequency of 297 Hz, and as a finite element model with a resonant frequency of 305 Hz. The device was then

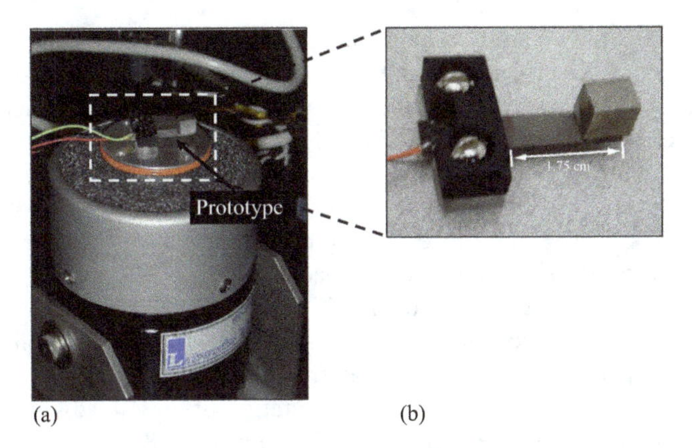


*Figure 3.9 (a) An assembled PEH developed by Roundy and Wright [128] mounted on a vibration shaker and (b) magnification of the harvester device*

characterized on a vibration shaker for a frequency sweep from 300 to 320 Hz across a resistive load of 2 kΩ–1 MΩ, and a shift to higher-frequency states with an increase in load resistance has been reported. The harvester is capable of producing 2 mW power from knee joint movements at a normal walking speed. A maximum voltage of 2.7 mV at an input frequency of 300 Hz across 2 Ω and 24 V at 320 Hz, across 1 MΩ load has also been recorded. However, the highest power was obtained across a 10 kΩ resistive load connected across the PZT plate of the harvester. Across a 20 kΩ resistive load at 310 Hz, a maximum voltage of 6.1 V is obtainable, resulting in an energy generation of 59 $\mu$J.

Roundy and Wright [128] reported a miniature (1 $cm^3$) PEEH, as shown in Figure 3.9(a). The prototype is capable of harvesting power from targeted machine vibrations of varying amplitudes, ranging from 0.02 to 1.01 g and frequencies ranging from 60 to 200 Hz. The harvester was tested at 120 Hz and under 0.25 g, it produced a power of 375 $\mu$W across a resistive load and 190 $\mu$W power across a capacitive load. Moreover, under the same base acceleration and input frequency, a 1.9 GHz radio transmitter may be operated by the devised harvester.

A novel PEEH has been designed and fabricated in [16], intended to be a power source for low-power microsystems, to make it autonomous. The harvester was fabricated by depositing AlN on the Si-on-insulator (SOI) wafer. An Al layer was deposited on AlN by the DRIE process to make contact and a top electrode. The highly doped top SOI layer was used as the bottom electrode of the PE material. A MEMS-scale device was developed and tested at a frequency of 900 Hz, under 1.02 g base excitation. It has been shown that the harvester is capable of producing 10 times more power with PZT (600 nW) as compared to AlN (60 nW) as a PE material.

The wing motion of a Green June beetle (*Cotinis nitida*) during the flight was studied by mounting a nonresonant microharvester on insects during flight [129]. The device uses low-frequency oscillations of the insect's wings motion to generate electrical power. PZT-5A, bimorph beams were fabricated on the Si substrate using DRIE and RIE techniques for prototype-I and prototype-II, respectively, with the final prototype with two spiral beams, fabricated using a femtosecond laser placed over the thorax of the insect. It has been shown that at 100 Hz, 18.5–22.5 $\mu$W can be generated from the wing stroke of the beetle during flight. The generated power from flight kinetics increases from 7.5 to 11.5 $\mu$W, with an increase in device size from 5.6 to 11 mm$^3$. Mounting both the generators, as shown in Figure 3.10(a) (one on each wing), can produce an output power of more than 45 $\mu$W. A total power of approximately 115 $\mu$W and a load voltage of 874 mV have been reported across 66.6 k$\Omega$ and 945 mV across the 200 k$\Omega$ load.

A nonresonant cantilever-based PEEH, as depicted in Figure 3.11, was fabricated to harvest traffic-induced bridge vibrations [130]. The bimorph steel plate, enclosed in upper and lower PZT plates, carries a tip mass weighing 12 g. Experimental results have shown that for excitation of only 0.02 g at 14.5 Hz, the potential difference across PE plates varied between 1.8 and 3.6 V, giving an average power of 0.03 mW.

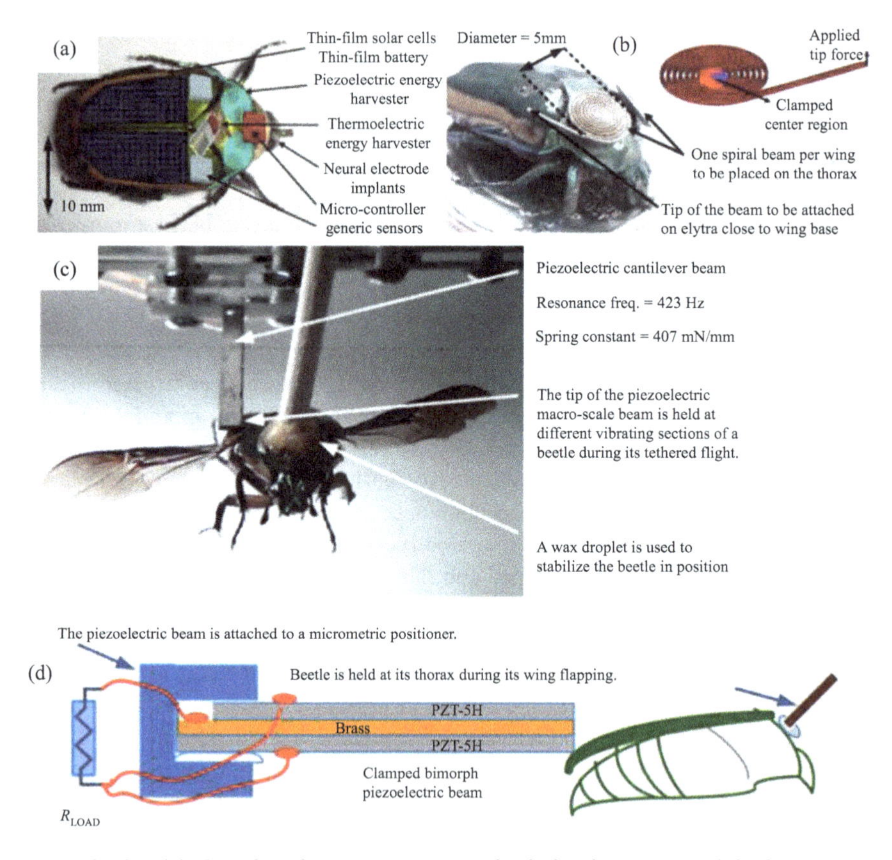


*Figure 3.10 (a) Graphical representation of a hybrid insect model, (b) spiral beam-type prototype placed on the beetle's thorax, (c) setup for in-lab power measurement during flight, and (d) PE cantilever beam for prototype-I and prototype-II [129]*

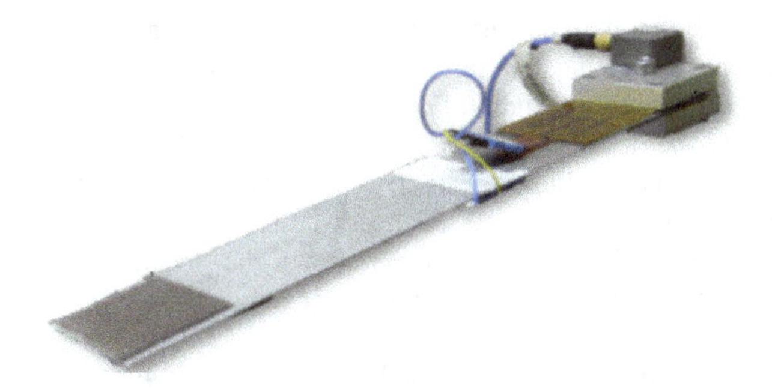

*Figure 3.11 Device architecture reported by Peigney and Siegert [130]*

*Figure 3.12 (a) Cross-section of the generator and (b) an illustration of the harvester implementation [133]*

### *3.2.2 Electromagnetic energy harvesters*

EMEHs work on the principle of Faraday's law of EM induction. As the conductor cuts magnetic flux in a closed circuit, a voltage is induced in the circuit [131]. The EM induction relates electromotive force ($V_{\mathrm{emf}}$) to the rate of change in magnetic flux and can be expressed as in (3.3).

$$V_{\mathrm{emf}} = -N \frac{d}{dt} \iint_S B \cdot dS \qquad (3.3)$$

where $N$ represents the number of turns in a coil, $B$ is the magnetic flux density passing through an area $dS$, and $V_{\mathrm{emf}}$ is the electromotive force. An EMEH commonly consists of a permanent magnet and coil arrangement, with either the magnet or the coil or both, attached to an oscillatory or a rotary structure for relative movement. An EMEH converts rotational, oscillatory, or linear motion into electrical energy. Generally, resonant-type EMEHs are more efficient, in terms of power generation and that is why most of the reported EMEHs are of the resonant type [132].

An EMEH, shown in Figure 3.12, which is capable of powering a bridge health monitoring sensor, has been fabricated and tested in [133]. The device has an overall volume of 68 cm$^3$. The casting and support in the harvester were fabricated from Al sections, with tungsten carbide proof mass used on its cantilever beam. Spiral and coils, made of Cu and NdFeB permanent magnets with N42 grade, were used to fabricate the device. At a low input frequency of 2 Hz and 0.055 g acceleration, the EMEH is capable of delivering an average power of 2.3 $\mu$W across a resistive load of 1.5 $\Omega$, with 57 $\mu$W peak power.

On testing over a bridge, it has been shown that the device can produce an average power of 0.5–0.75 $\mu$W for a frequency range of 2–30 Hz at 0.5–1 g base acceleration amplitudes.

Two EMEHs have been developed to extract energy from railway track vibrations, as reported in [134]. The developed harvesters, shown in Figure 3.13(a), are capable of having bidirectional track vibrations conversion into unidirectional oscillations using a nonlinear mechanical motion rectifier mechanism to ensure higher energy density at low base excitations. When a train passes at a 30 km/h speed, prototype-I can produce 1.12 W average power, which is exactly half of the power produced by prototype-II power at 2.24 W. Prototype-I delivers 0.8 W power across a 2 $\Omega$ load resistance at 1 Hz at 1 mm displacement amplitude, while

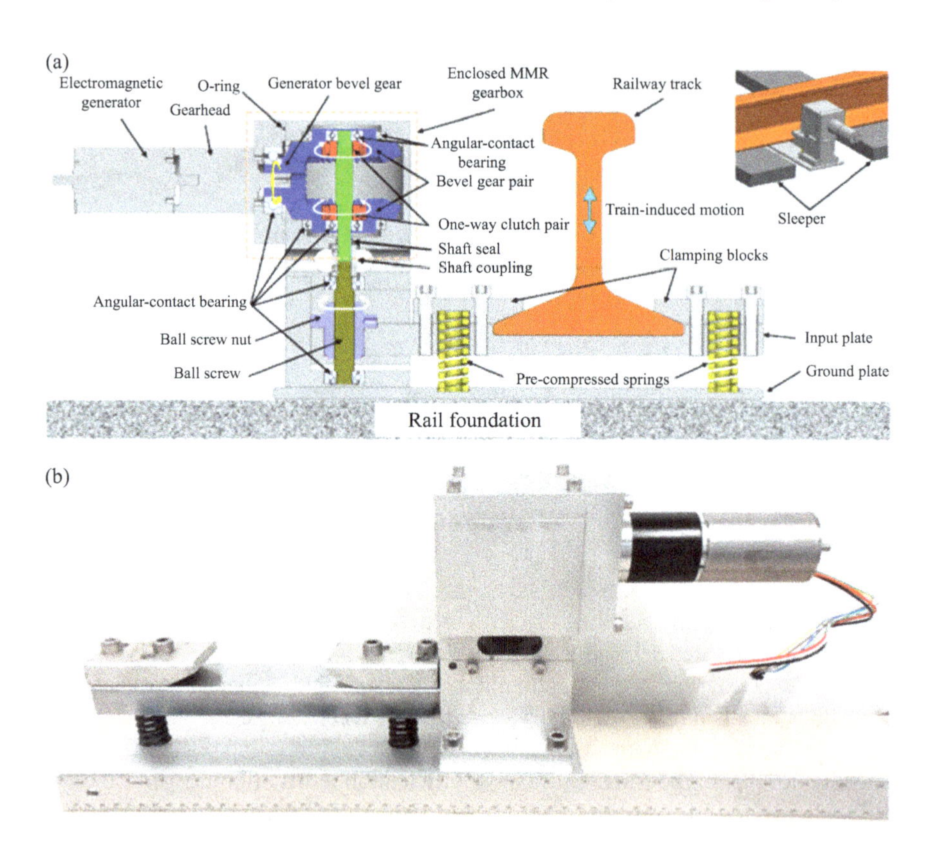


*Figure 3.13 (a) Railway track EH design and (b) prototype developed in [134]*

prototype-II generates 0.9 W across the same load. This is due to the higher electrical damping coefficient of prototype-II.

Khan and Iqbal [135] have developed a 2-DOF EMEH with coil and magnet being placed on different cantilever beams to allow a broad bandwidth at low-frequency oscillations. The length of the beams as well as the gap between the coil and magnet is adjustable depending on the amplitude of base excitations. The harvester has been characterized under different base accelerations from 0.2 g to 0.6 g, so as to produce a maximum power of 1 955 $\mu$W across a 28 Ω load under 0.6 g base excitation at a low resonant frequency of 7.6 Hz.

A bistable nonresonant EMEH with a size of 3.7 $mm^3$, which responds well at a low frequency of below 30 Hz oscillations, has been developed in [136]. The fabrication process steps involve the deposition of Cu alloy on a Si wafer, following ferrous chloride ($FeCl_2$) etching. The proof mass of the prototype and coils, as shown in Figure 3.14(b), was fabricated from Cu and a spring made of Cu alloy carried the permanent magnets. Upon exposure to base excitations, the harvester is capable of producing an average power of 39.45 $\mu$W, with a peak power of 558.2 $\mu$W, across a resistive load of 270 Ω at 10 Hz at 1 g base acceleration with an average power density of 10.7 $\mu W/cm^3$.

A biomechanical EMEH with rolling magnets is presented in [137]. An 8 cm × 2 cm plastic channel encloses the identically sized two cylindrical magnets, with a coil having 2 kΩ optimum load resistance, wrapped around the plastic channel, as shown in Figure 3.15(b). The harvester was tested under the shaker's vibrations: during walking, running, handshaking, and cycling. Results have shown that sufficient power to operate wearable sensors autonomously is produced by the EMEH. At 4 km/h normal walking speed, the harvester is capable of producing 500 $\mu$W power, while from handshaking, it can generate an average power of 1.02 mW at an input frequency of 3.1 Hz under 1 g base acceleration. Experimental results

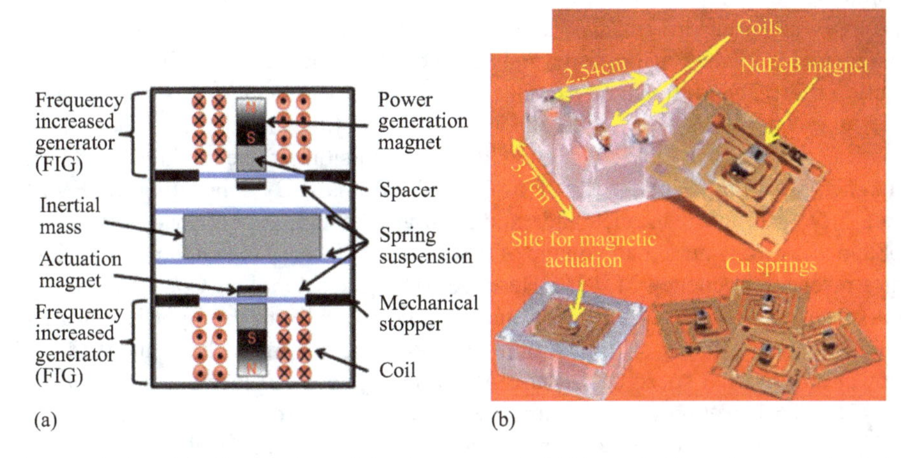


*Figure 3.14 (a) Illustration of the harvester in [136] and (b) a photograph of the assembled device*

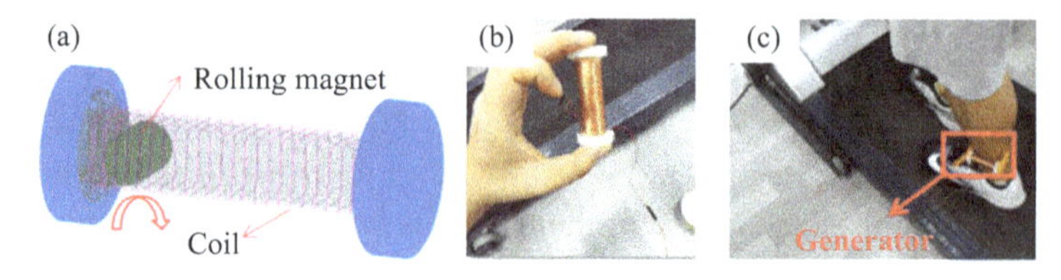


*Figure 3.15 (a) Schematic of the proposed harvester, (b) handshaking experiment, and (c) walking on a treadmill [137]*

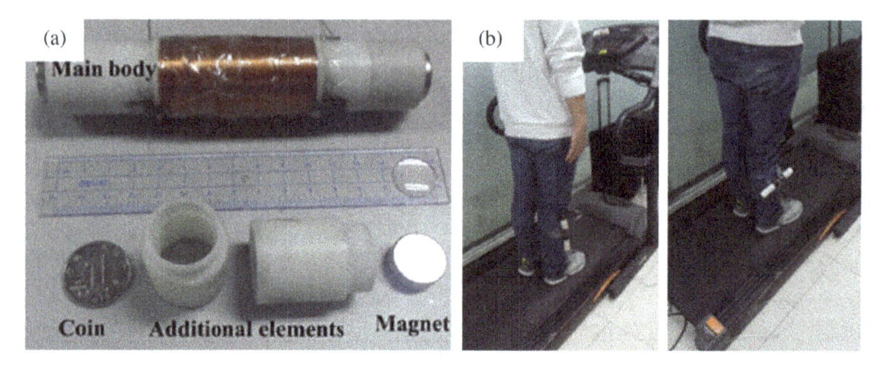


*Figure 3.16 (a) Device assembly and (b) application to the human body [138]*

have shown that the harvester generates 1.1 V, allowing 2 min operations of a microsensor, from about 1-min handshaking with the same frequency and amplitude. Moreover, 1 min of cycling can generate 220 $\mu$W power.

A human motion-based EMEH has been developed in [138]. The device is comprised of a 25 mm long, 20 mm diameter, and 4 mm thick hollow tube that encloses a permanent magnet, as shown in Figure 3.16(a). Two identical-sized magnets are located at opposite ends of the tube. The tube was fabricated from Teflon using 3D printing, and a circular coil containing 480 turns was wrapped over the tube. The harvester weighs 103.7 g. On subjecting the EMEH to body kinematics, the magnets started moving due to repulsion of the middle cylindrical magnet, resulting in induced EMF at the coil terminals. By connecting the EMEH to a 5 Ω optimum resistance and exciting it at 0.35 g, a 0.34 V load voltage was obtained. A maximum load voltage of 0.86 V was delivered to matching impedance at a resonant frequency of 9.155 Hz under 0.85 g acceleration. Consequently, the harvester was connected vertically and then horizontally to the right leg of a test subject during walking on a treadmill, generating a 224.5 mV voltage and 0.42 mW power at a normal speed of 5 km/h. At a running speed of 9 km/h, it produced a 460 V and 6.1 mW power at the time the leg was on the ground. A maximum power of 10.6 mW was generated at 8 km/h when it was fused transversely to the leg. The authors concluded that the generated power can be further increased with increasing walking speed and by enhancing the number of turns in the coil.

The EMEH, as reported in [139], can power health-monitoring body-worn sensor nodes. The EH produces an electrical output by the EM coupling of a translator and a stator. The translator serves as proof mass and initiates movement in the harvester, with bearing working as elastic spring, an insulator between the stator and translator and guide for the stator motion. Two rods were used to attach the bearing to the translator and the coil is connected to a PCB circuit. With the help of small screws, a provision is made to allow gap adjustment between the coil and the harvester housing, as needed. The overall assembled structure weighs 43 g and occupies a volume of 0.49 cm$^3$, having an internal resistance of 5 k$\Omega$. The prototype can be seen in Figure 3.17(c) and (d) during lab testing, mounted on a vibration shaker with sinusoidal excitations, and worn by a test subject (human), respectively. Mounted on an upper arm, it can generate 5.1 $\mu$W power across a 10 k$\Omega$ load. A power of 22.0 $\mu$W was reported by the authors, at a normal walking speed when the EMEH is worn on a knee joint, which is sufficient for the operation of most body sensors. The device produces a voltage in the range of 0.5–4 V and a power of 5–25 $\mu$W across various load resistances.

The effect of magnetic repulsion on the natural frequency and bandwidth of a vibration-induced cantilever-type EMEH is investigated in [140]. Due to the magnetic repulsion of like magnetic poles, a significant decrease in the natural frequency of the cantilever beam has been observed with end magnet vertically oriented, as shown in Figure 3.18. Experimental results have shown that for every increase of 0.1 g base acceleration, an increase of 0.9 Hz in natural frequency was recorded when both the magnets were oriented horizontally and 0.6 Hz in case of vertical configuration.

In [141], an EMEH with an optimized guided beam has been shown to produce a maximum power of 5.4 mW when exposed to mechanical vibrations. The harvester comprises two cylindrical magnets and Cu wound coils and generates power as a result of repulsive forces experienced by the adjacent magnets during excitation.

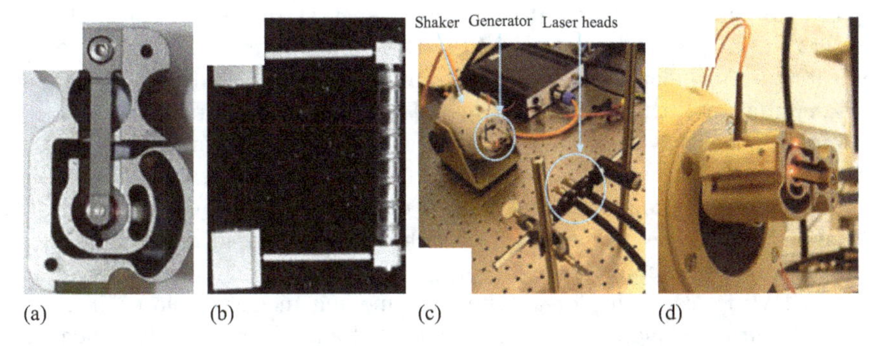


*Figure 3.17 The fabricated device reported in [139]: (a) open generator top view, (b) translator and bearing, (c) experimental setup, and (d) translator motion indicated by red laser light*

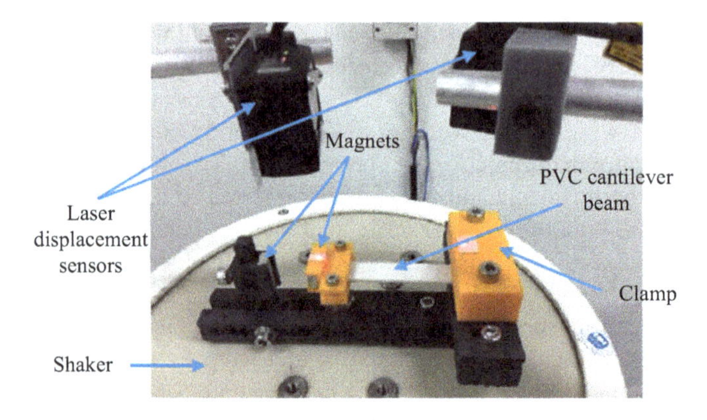


*Figure 3.18 The fabricated device reported in [140] under test*

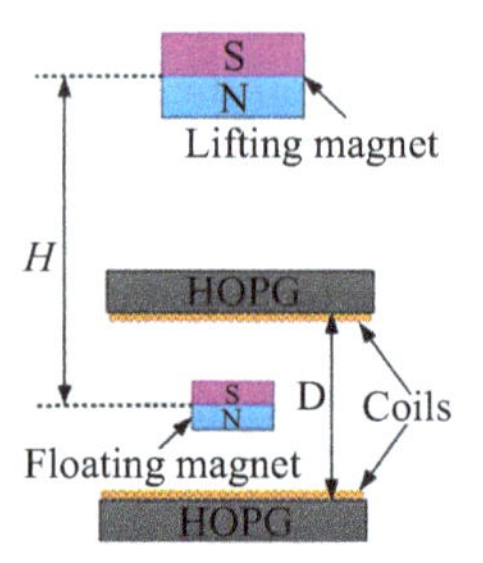


*Figure 3.19 Schematic of the EMEH in [142]*

An EMEH of low-frequency oscillations with a 2.8 Hz resonant frequency is presented in [142]. The harvester consists of two small disk-shaped permanent magnets and two pyrolytic graphite sheets holding four wound coils, with 120 turns in each coil, above and below the moving magnet, as shown in Figure 3.19. The coil has an internal resistance of 27 Ω. The prototype can be installed in civil infrastructure including bridges and tall buildings, to harvest low-frequency and low-amplitude oscillations. The authors in [142] have characterized the device for forward frequency sweep from 0.4 to 4 Hz and a base excitation range from 0.1 to 1.0 g. When the gap between the coil and magnet was kept at 2.52 mm, the harvester successfully produced a peak voltage of 268 mV and an average power of 62 $\mu$W at resonance. The output power of the harvester can be increased by reducing the gap between the coil and the magnet.

An EMEH, consisting of a cylindrical moving magnet enclosed in a tube, wrapped with a wound coil, has been fabricated and experimentally validated in [143]. Two magnets are fixed at both ends of the tube, to produce a repulsive force on the central moving magnet, which kept it floating inside the tube and ensured magnetic induction across the wound coil. Figure 3.20 depicts the EMEH.

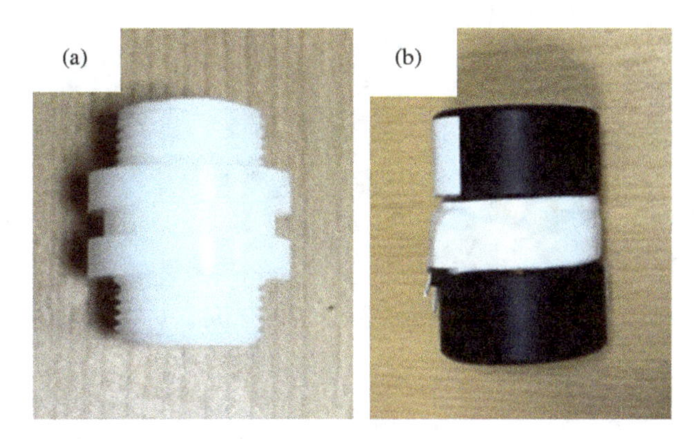


*Figure 3.20 (a) Fabricated hollow tube and (b) harvester with fixed magnets housing and coil [143]*

Experimentally, the harvester delivered a 7.7 mW power and a load voltage of 1.8 V across a 220 Ω when excited at 8.9 Hz under 0.24 g acceleration.

### 3.2.3 Hybrid energy harvesters

To achieve a more energy-efficient system with enhanced power-producing capabilities, multiresonant states, and wide operation frequency bandwidths, researchers have been looking into combining more than one energy harvesting techniques into one system, i.e., HEH. This hybrid approach to energy harvesting, may either be an integrated system with different energy harvesting systems, that is a single system with more than one generator using different conversion mechanisms or a single system with multiple energy harvesting sources. Various energy conversion technologies such as PE, EM, triboelectric, electrostatic, and acoustic may be combined in a hybrid harvesting system.

On its own, the PEEH conversion mechanism is considered as the best harvesting technologies as higher power densities may be obtained from the same vibration source. Combining with the EM conversion mechanism, the total power of the vibration-based HEH which combines PE and EM mechanisms is a function of the total damping of the system, with an increase in the system damping decreasing the vibration amplitude of the EM mechanism [144]. Recently, hybrid nano EHs, combining PE and triboelectric materials, have attracted considerable attention from researchers and may provide a solution into making WSNs as power autonomous systems, by taking advantage of both PE and triboelectric effects [5,145–150].

A HEH, integrated onto a computer keyboard, is reported in [151]. The device comprises two Ni electrodes, a PZT layer, and a spiral planar coil, as shown in Figure 3.21(a). The coil was fabricated by depositing parylene in a vacuum and patterned by RIE to expose the contact pads used as electrodes. By sputtering and patterned in the wet etchant, Al was deposited and insulated by depositing a second parylene layer. Finally, the upper electrode was protected with parylene

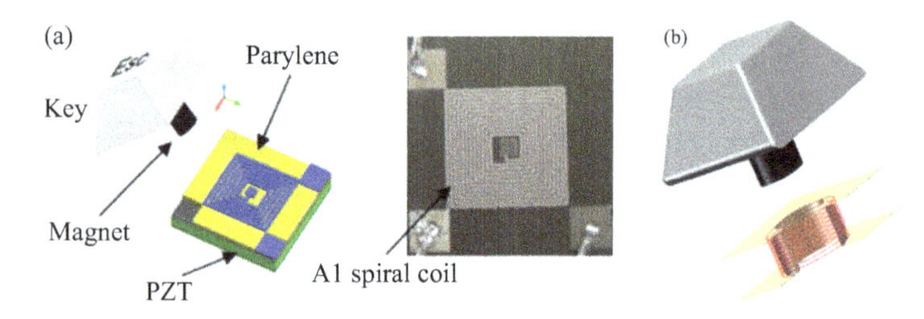


*Figure 3.21 (a) 3D view of the fabricated device and (b) a wound copper coil [151]*

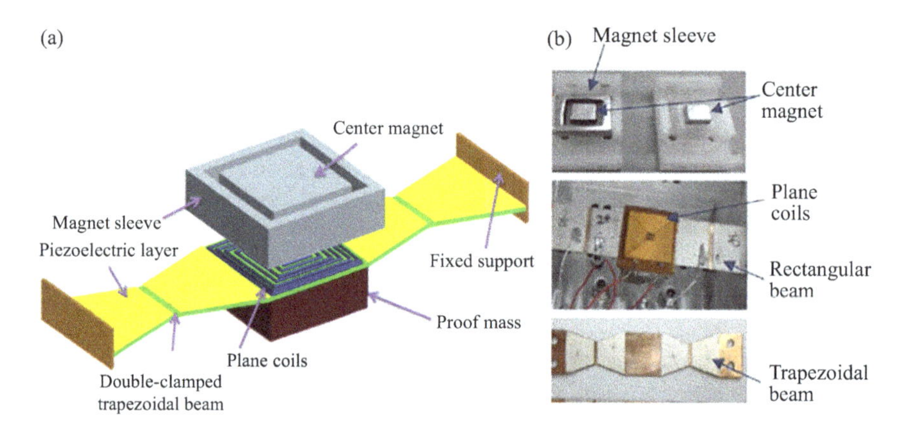


*Figure 3.22 (a) Proposed HEH in [152] and (b) the developed hybrid harvester*

protective passivated coating. The hybrid harvester was tested by putting the HEH onto a keyboard, to see the amount of power that may be harvested during normal typing. A maximum PE power of about 40.8 $\mu$W across a 3 M$\Omega$ load resistance and an EM power of 1.15 $\mu$W across a load of 35 $\Omega$ were produced during the experiment. It has been shown that the harvester can produce 3.46 mW of power, which can be used to recharge a battery or to operate a small circuitry while the keyboard is in use.

A low-frequency HEH with a double-sided PE trapezoidal beam holding three printed planar coils encapsulated in a proof mass and magnet has been developed in [152] and is shown in Figure 3.22. The magnet is wrapped in a sleeve to narrow down magnetic field lines, ensuring better magnetic induction. The trapezoidal PZT beam is capable of generating more uniform strain than the conventional rectangular beam in a harvester. Under low base excitations of 0.2 g, the hybrid PE-EM harvester produces 637 $\mu$W power, which is an overall increase of 52.4% as compared to power produced from the stand-alone PE or EM power conversion system.

A broadband hybrid EM-TEEH has been developed in [153]. A poly-dimethylsiloxane (PDMS) spring, as shown in Figure 3.23, was fabricated from

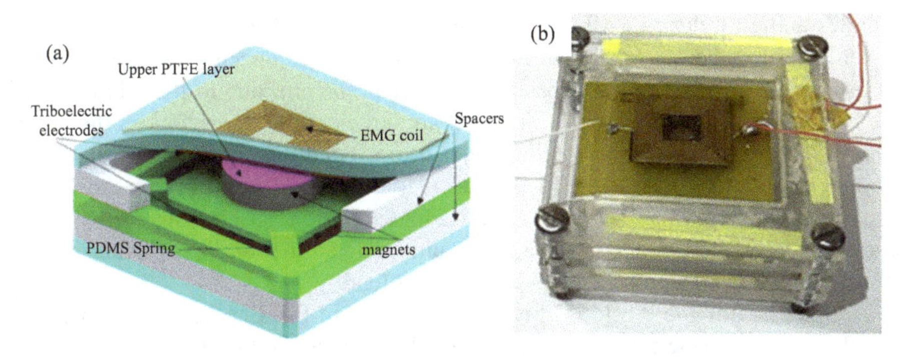


*Figure 3.23 (a) Schematic of the hybrid harvester reported in [153] and (b) the fabricated device image*

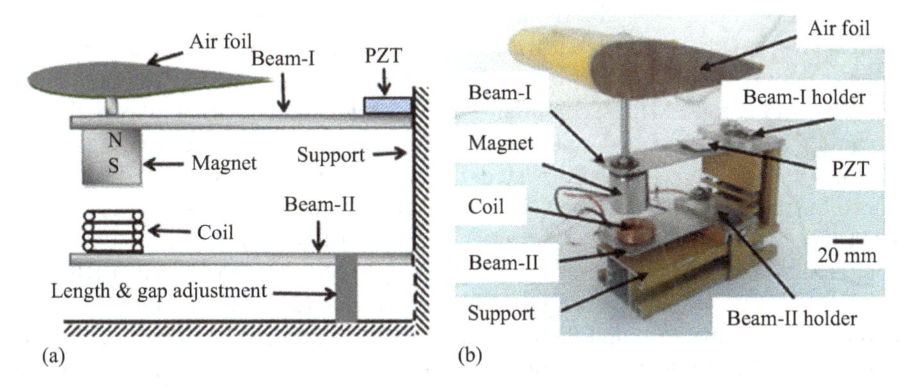


*Figure 3.24 (a) Schematic of the HEH and (b) a photograph of the fabricated bridge EH. Reproduced with permission from [154].*

metallic mold, prepared from SUS304 steel. The mold was filled with PDMS, which was then peeled off from the mold to obtain a resonating four-arm PDMS spring. The internal impedances of EM and triboelectric parts of the fabricated EH were found to be 15 Ω and 75 MΩ, respectively. When tested under a base acceleration of 1.5 g at their resonant frequencies, 35.27 and 0.166 $\mu$W power were produced, respectively, across the matching loads. The multimodal harvester was also tested under base acceleration from 0.1 to 2 g and it has been shown that 0.2 $\mu$W power was produced at a resonant frequency of 80 Hz under 2 g. The generated output of the harvester corresponds to 0.8 $\mu W/cm^3$.

A HEH, based on PE–EM conversion mechanisms and designed for bridge health-monitoring applications, is reported in [154], as shown in Figure 3.24. The harvester is capable of generating electrical energy by harvesting mechanical vibrations and wind simultaneously using PE and EM conversions when exposed to traffic-induced bridge vibrations and ambient wind. When subjected to external vibrations, the upper cantilever beam holding the cylindrical magnet as the tip mass vibrates and an induced EMF is generated across the coil terminals placed over a lower cantilever beam. At the same time, PE power is produced by piezoelectricity across the PZT plate, which is pasted on the upper beam, due to the vibrations of the beam. Similarly, when traffic-induced air surges strike the airfoil on the upper beam, the upward lift force oscillates the beam, resulting in an electrical output from both mechanisms. The EM portion of the reported harvester is capable of generating 2 214 $\mu$W power across a 28 Ω matching load at a resonance of 11 Hz under a base excitation of 0.6 g, while an output power of 156 $\mu$W is produced from PE part under 0.4 g across a 130 kΩ matching load.

Rajarathinam and Ali [155] investigated a hybrid PE–EM EH, consisting of a cantilever beam made of mild steel carrying the tip mass and PZT plate attached on its top side. A magnet, acting as a proof mass, is attached to the bottom side of the beam to generate an induced EMF in the coil when the device is excited mechanically. When exposed to mechanical excitations, the harvester produces EM power across the coil and PE output from the PZT material placed over the deflected beam. It has been shown that the harvester operates over a broad frequency with better power generation due to the coupled PE–EM effect.

Hybrid PE and EMEHs, comprising a PE cantilever made of multilayered PZT, a pair of magnets and a double-layered printed copper coil, are investigated in [156]. The 3-D coils were fabricated by depositing plasma-enhanced chemical vapor on the substrate and then, insulated with an undoped silicon glass (USG) layer. For the fabrication of the top and bottom electrodes, 1 $\mu$m thick Al layer was sputtered above the USG layer, followed by $Si_3Ni_4$ layer deposition and RIE patterning to expose the connecting metal. Moreover, a 10 $\mu$m thick USG layer was deposited and patterned, and the coil was then electroplated. Each turn of the coil was separated from the adjacent one by the chemical mechanical polishing process and the coil's electrodes were patterned, as shown in Figure 3.25(b). Under 2.5 g base acceleration and at 310 Hz, the PE portion produces an RMS voltage of 0.84 V and a power of 176 $\mu$W that corresponds to a power density of 790 $\mu W/cm^3$. Moreover, the EM portion of the hybrid harvester generates 0.78 V and 0.19 $\mu$W with a power density of 0.85 $\mu W/cm^3$, under the same acceleration and resonant frequency.

Another hybrid PEM-IEH is reported in [157], which consists of a compact, three degree of freedom system with an upper and lower PE cantilever beam holding wound coils. An intermediate steel beam holds two magnets on its top and bottom sides as proof mass. The mm scale (30 mm × 30 mm, including the casing) harvester, shown in Figure 3.26, can be fixed inside the sole of a shoe and be a part of the sole. The harvester exhibits multiresonant frequencies and wide operating bandwidths. At a resonant frequency of 9 Hz, it is capable of delivering 95 $\mu$W power to a matching impedance of 14 Ω under 0.6 g.

Hybrid footwear EH based on combined triboelectric, EM, and PE mechanisms is presented in [158]. The triboelectric part of the harvester is composed of three optimized pairs of parallel plates, while the EM part is composed of a coil (27 mm × 22 mm) and a magnet (4 cm × 4 cm), which is separated by an 8 mm thick sponge. ZnO nanowires and ZnO thin films, which have been prepared by

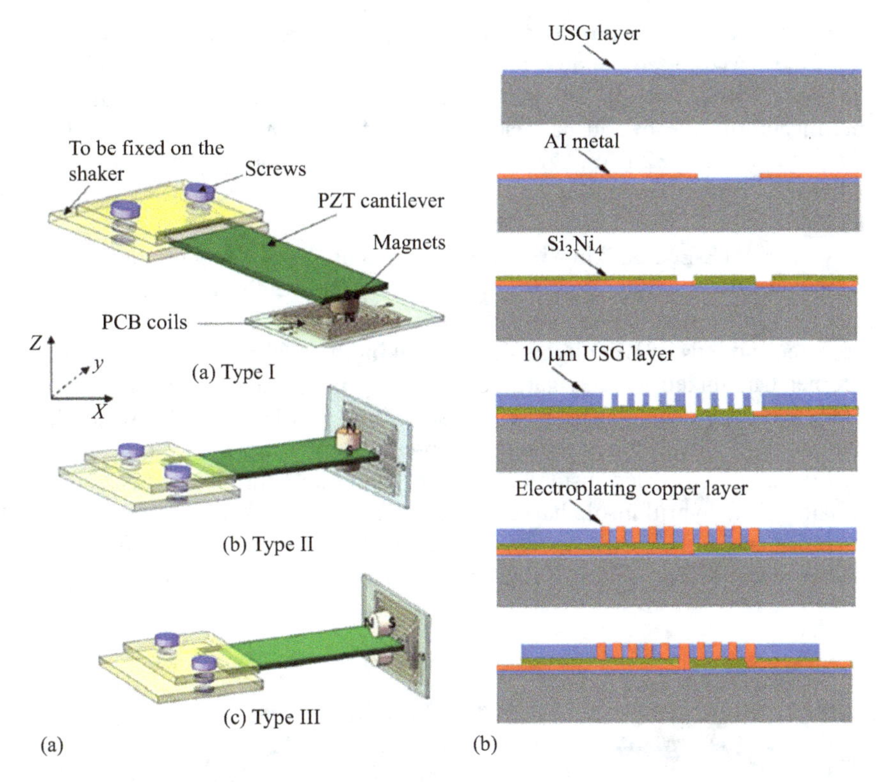


*Figure 3.25 (a) Three prototypes with different assemblies with respect to magnets and (b) the process flow of coil fabrication [156]*

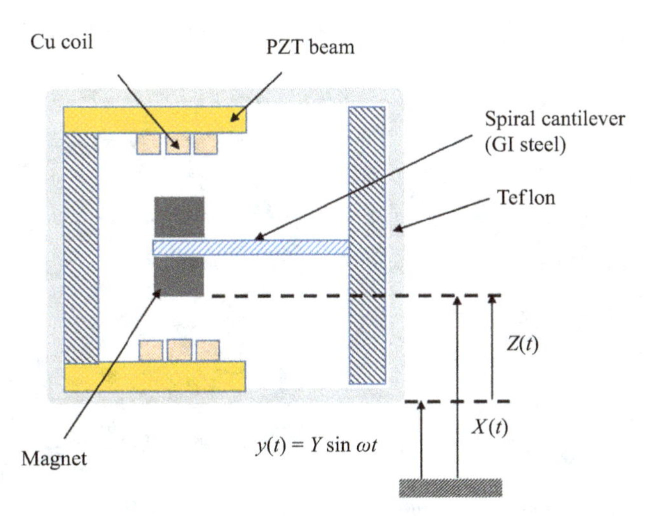


*Figure 3.26 Cross-section of the insole EH. Reused with permission from [157].*

hydrothermal and ion beam deposition processes, respectively, constitute the PE part of the harvester. As shown in Figure 3.27, the EM part is placed above the triboelectric part, with the PE part kept on top of the harvester. During in-lab experimentation, the hybrid harvester generated an open-circuit voltage of 75 V and a peak power of 32 mW. Upon integration into the shoe, the harvester was able to charge a 220 $\mu$F capacitor up to 1 V and light up three LED lights during walking.

A hybrid TE-EM walking EH, shown in Figure 3.28, for providing power to wearable electronics, is presented in [159]. The light-weight, cm-scale (62.5 cm$^3$) device was integrated into the sole of a commercial shoe and was used to power a pedometer and tens of LED lights during walking. For fabrication, a liquid PDMS elastomer was mixed with Sylgard 184, Dow Corning cross-linker and then a triboelectric charge plate was prepared by depositing PDMS on the Si pyramid. The mixture was moved to the Si template, spin-coated for 1 min at 500 rpm speed and the Si template was dried inside an oven for about an hour. The triboelectric and EM parts of the hybrid insole harvester can produce 4.9 mW power across a 6 M$\Omega$ load resistance and 3.5 mW across a 2 k$\Omega$ load, respectively. In terms of power densities, the triboelectric and EM parts of the harvester produce 5.1 and 3.6 W/m$^2$,

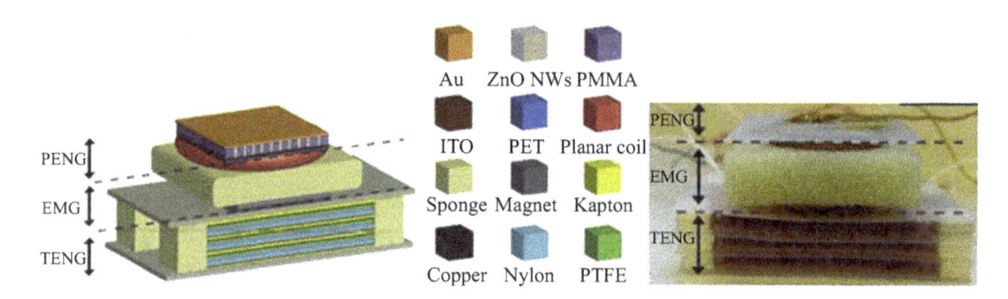


*Figure 3.27 (a) Schematic and (b) photograph of the fabricated hybrid harvester [158]*

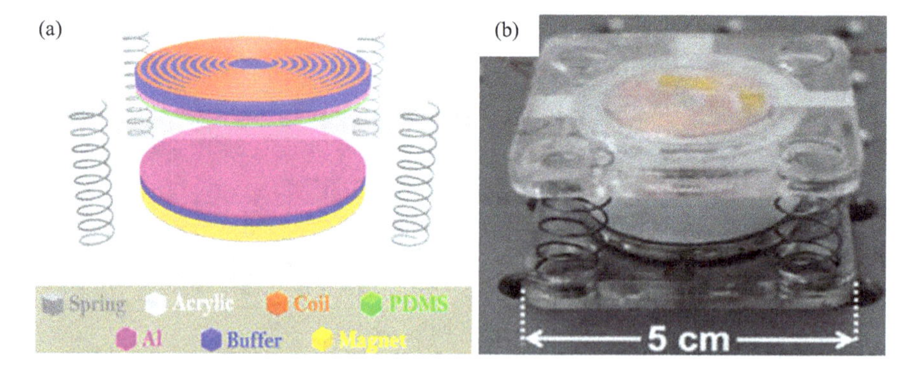


*Figure 3.28 (a) Schematic of the hybrid harvester reported in [159] and (b) a photograph of the fabricated prototype*

respectively. Upon integration into a shoe, it has been shown that the harvester can charge a Li-ion battery from 0.5 to 0.8 V, with 5 min of normal walking.

A multimodal hybrid PEM-IEH has been developed in [160], as shown in Figure 3.29. The prototype is comprised of an encapsulated spiral spring between two PVDF stretchable strips. The spring is holding an upper and lower magnet at its central platform, while wound coils have been attached to the piezoelectric strips (PVDF) to ensure dual transductions on exposure to foot excitations. The upper and lower unit of the harvester generated a load voltage of 3.34 and 3.83 V, respectively, on connecting across 9 MΩ load resistance, which is equal to the internal impedance of the PVDF. A 100 μF capacitor was charged up to 2.9 V as a result of 8 min normal walking. The combined harvesting units showed 30% increased voltage than that of the individual units.

A nonlinear hybrid PE–EM EH is reported in [161]. The designed harvester is composed of three pairs of cylindrical magnets in line with each other, as shown in Figure 3.30(b). The upper and lower magnets are fixed, equidistant from the central magnet and encircled by wound coils. Movement of the central magnet results in an

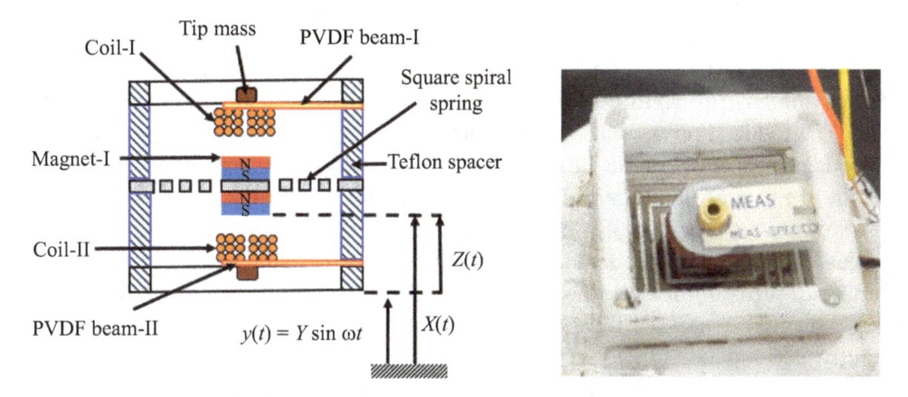


*Figure 3.29 A hybrid insole energy harvester (HIEH): (a) vertical cross-section of the HIEH and (b) top view of the assembled HIEH*

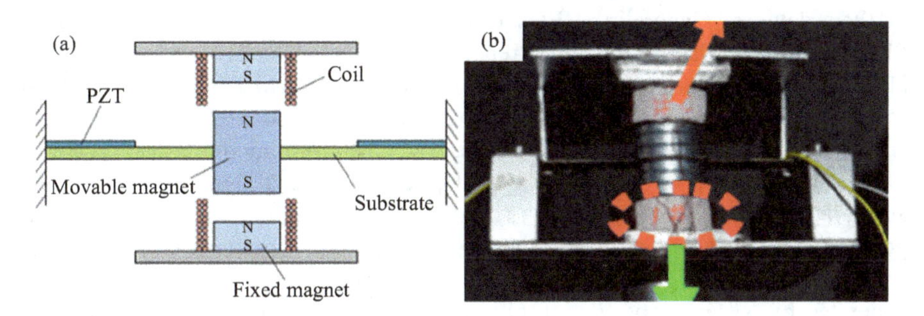


*Figure 3.30 (a) Illustration of the nonlinear HEH proposed in [161] and (b) fabricated harvester*

induced EMF, due to the relative motion of the intermediate magnet. The PE material layers were pasted on the top side of a fixed beam, with the beam being enclosed within the central magnets. The PE layers added power to the hybrid harvester, following the piezoelectricity principle when the magnetic mass vibrates. Under 0.3, 0.45, and 0.5 g acceleration levels, the hybrid harvester generates 0.52, 1.76, and 3.54 mW power, respectively, across the load resistance from the EM part. With an increase in base excitation from 0.3 to 0.6 g, the resonance decreases to 110.5 Hz from 113.2 Hz. Under 0.2 and 0.45 g, the PE portion harvests 0.085 and 0.5 mW power across 140 and 190 kΩ, respectively. Moreover, it has been reported that the optimum power generated from the EM part of the hybrid device under 0.2 and 0.4 g are 0.14 and 1.19 mW, respectively.

## 3.3 Comparison and discussion

A comprehensive comparison of VEHs reported in the literature is shown in Table 3.4. The harvesters are assessed based on important parameters, such as the resonant frequency, base acceleration, harvester's dimensions, internal resistance, generated voltage, harvester's power, and power density. For PEEHs, PZT is the most beneficial among the PE materials, due to low hysteresis losses, non-centrosymmetric crystalline structure, high power density, and good signal-to-noise ratio. The induced power in EMEHs is directly reliant on the number of turns in the coil, magnet size, magnetic flux density, the gap between the coil and the magnet, weight of the proof mass, and the material used for beam and device size.

Power as a function of the harvester's dimension is given in Figure 3.31. An increase in the overall dimensions of the harvester including the magnet size, the number of turns in the coil, the thickness of the PE material and cantilever beam's size, generally increases the power of the harvester. Similar conclusions can be directly derived in [135,154,159,172,179,180], with these harvesters capable of delivering more power due to their comparative greater sizes. Less power is produced from the relatively small-sized harvesters as reported in [16,176,168,173]. However, the reported harvesters in [156,183] can generate more power despite their small dimensions by implementing both PE and EM conversion mechanisms in their devices. Generally, HEHs can generate more power than stand-alone PEEHs and EMEHs. Only the reported PEEHs in [171] and EMEHs in [135,179,180] produce comparable power levels to hybrid EHs; however, these may be largely attributed to their larger sizes. The sizes of PEEHs are generally smaller and have a greater tendency to be developed into micro and nanoscales.

Figure 3.32 shows maximum output power produced as a function of base acceleration for the reported EHs. As base excitation increases, the resultant output power generally increases due to the higher amplitude vibrations experienced by the device. A similar generalization is given in [156]. Comparatively, less power is generated by Rezaeisaray *et al.* [168] and Galchev *et al.* [181], as these devices were subjected to low base excitations. Harvesters reported in [154,157,161,184] produce high power than most of the harvesters at even low base accelerations

*Table 3.4 Comparison of the reported EHs*

| Mechanism | Material | Device size ($cm^3$) | Resonant frequency (Hz) | Acceleration (g) | Internal impedance (Ω) | Voltage (mV) | Power (μW) | Power density (μW/$cm^3$) | Ref. |
|---|---|---|---|---|---|---|---|---|---|
| Piezoelectric | PZT-5A | 86.4 | 260 | 0.5 | 20k | 2.47 | 51 | 0.59 | [122] |
| | PZT | — | 20 | 0.8 | — | — | 61.5 | — | [123] |
| | PZT and Al | 0.072 | — | — | 400k | 12 | 800 | 11 111 | [124] |
| | PZT-PIC 255 | — | 2 939 | — | 68k | — | 660 | — | [126] |
| | PZT 8% PT | | | | 215k | | 571 | | |
| | PZT-5H | 0.00845 | 300–320 | 0.72 | 0.0002–1 000 | 2.7–24 | 59 | 6 982.24 | [127] |
| | PZT-5A | — | 120 | 0.255 | — | 1.2 | 375 | — | [128] |
| | PZT and steel | 0.2 | 900 | 1.02 | — | — | 0.06, 0.60 | 7.5, 75 | [16] |
| | PZT | — | 14.5 | 0.02 | 100 | 1.8–3.6 | 30 | — | [130] |
| | PZT-5H | $8 \times 10^{-9}$ | 80 | — | 333 | 1.2 | 2 | $250 \times 10^6$ | [162] |
| | PZT | — | 13 900 | — | 5 200 | 2.4 | 1 | — | [163] |
| | PZT | — | 60.3–1 060.6 | — | — | 0.51–7.45 | — | — | [164] |
| | AlN | 1.47 | 53.8 | 0.9 | 24k | 20.57 | 4 500 | 3 051 | [165] |
| | PZT, silicon, and brass | 5.3 | 56 | 2 | 340k | 1.7 | 2 120 | 400 | [166] |
| | PZT-5A | 12 | 246 | — | — | 2.51 | — | — | [167] |
| | PZT and AlN | 0.004 | 71.8–188.4 | 0.2 | 2 M | 1 | 0.136 | 34 | [168] |
| | PZT and steel | — | 12.4 | 1 | 95 | 7.2 | 800 | — | [169] |
| | PZT, Al, and Cu | — | 10–30 | 1 | 7.1 M | 13.2 | 3.62 | — | [170] |
| | Al, steel, and PZT | 2.048 | 5 | — | 400k | 556 | 2 100 | 1 025.39 | [171] |
| | PZT, Cu, Al, and nylon | 2 407.98 | 5 | 0.05 | 400k | — | 26.6–30.1 | 0.01 | [172] |
| | PZT | 0.01 | 30–47 | 1 | 330k | — | 0.0855 | 8.55 | [173] |
| Electromagnetic | Cu and steel | 38 | 16 | — | 8 | 750 | 14 000 | 368.4 | [116] |
| | Cu and Al | 68 | 2 | 0.05 | 1 500 | 375 | 2.3, 57 | 0.83 | [133] |
| | Cu and Al | 192 | 44 | 0.6 | 28 | 385 | 1 955 | 10.18 | [135] |

(Continues)

*Table 3.4 (Continued)*

| Mechanism | Material | Device size ($cm^3$) | Resonant frequency (Hz) | Acceleration (g) | Internal impedance (Ω) | Voltage (mV) | Power (μW) | Power density ($\mu W/cm^3$) | Ref. |
|---|---|---|---|---|---|---|---|---|---|
| | Al and steel | 0.00507 | 10 | 1 | 270 | — | 39.45 | 778.01 | [136] |
| | Cu and plastic | | 3.1 | 1 | 2k | 1.1 | 1 020 | — | [137] |
| | Cu and Teflon | — | 5.5–7.2 | 0.35–0.85 | 5 | 340 | 6 010 | — | [138] |
| | PPVC, Latex, and Cu | 24.73 | — | — | 0.0043 | 1 200 | 18 600 | 752.12 | [174] |
| | AlN and Al | 0.49 | 12.4–16.01 | 0.5 | 10k | 500–4 000 | 5–25 | 51.01 | [139] |
| | Cu and ABS | — | 2.8 | 0.1–1 | 27 | 268 | 62 | — | [142] |
| | Cu and Teflon | — | 8.9 | 0.24 | 220 | 1 800 | 7 700 | — | [143] |
| | Cu and Al | 68 | 2–30 | 0.01–0.1 | 15k–40k | 0.015 | 2.3 | 0.033 | [175] |
| | AlN and PZT | 0.008 | 300 | 1.02 | — | — | 0.8 | 100 | [176] |
| | Al (EN AW-7010) | 3.75 | 10 | 1 | 220 | 0.1 | 22, 163 | 3.62 | [177] |
| | Cu, rubber, and Teflon | 5.83 | 27 | 1.02 | 3.6 | 13 | 2.1 | 0.36 | [178] |
| | Cu and Al | 1 143 | 2.2 | 0.01 | 465 | 700 | 1 380 | 1.20 | [179] |
| | Cu and Al | 700 | 3.6 | 0.2–0.4 | 54.5 | 206 | 354.5 | 0.506 | [180] |
| | Cu and Al | 43 | 2 | 0.5 | 1 500 | 0.2 | 2.3 | 98.9 | [181] |
| Hybrid | PZT, Al, and parylene | 0.0191 | — | — | 3 M, 35 | — | 40.8, 1.15 | 2 136.12 | [151] |
| | PDMS, steel, and Cu | — | 80 | 1.5 | 15, 79 M | — | 35.27, 0.166 | 0.8 | [153] |
| | PZT, Al, and Cu | 192 | 11–43 | 0.2–0.6 | 300k, 28 | 6 421, 483 | 155.7, 2 214 | 11.5 | [154] |
| | PZT, Al, and Cu | 0.222 | 310 | 2.5 | 3 M, 710 | 840, 0.78 | 176, 0.19 | 790, 0.85 | [156] |
| | Al, steel, and Teflon | 46.8 | 8, 25, 50, and 51 | 0.5 | 41, 43.5 | 160, 190 | 1 670 | 35.68 | [182] |

*(Continues)*

| | | | | | | | | |
|---|---|---|---|---|---|---|---|---|
| PVDF, steel, and Teflon | 33.3 | 7.5–40 | 0.1–0.6 | 3 M, 15 | 55 | 65 | 1.95 | [157] |
| Cu, Au, sponge, and ZnO | — | — | — | 1 M, 70 | 14 | 32 | | [158] |
| Cu, Al, acrylic, and buffer | 62.5 | — | — | 6 M, 2k | 300 | 4 900, 3 500 | 78.4, 56 | [159] |
| Cu, Teflon, and PVDF | 44.1 | 9.7, 41, 50, and 55 | 0.6 | 13.5, 16.5, and 9 M | 7,170 | 109, 179 | 4.05 | [160] |
| Cu, PZT, and steel | — | 110–119 | 0.2–0.6 | 190k, 21 | 9, 220 | 850–3 540 | — | [161] |
| Si, Cu, and PZT | 0.972 | 55.9–56.3 | 0.2 | 300k, 400 | 3.6 | 40.62 | 41.79 | [183] |
| Cu, Al, and PZT | — | 12–22 | 0.4 | 90k, 10 | 1.21, 12.5 | 244.1, 250.2 | — | [184] |

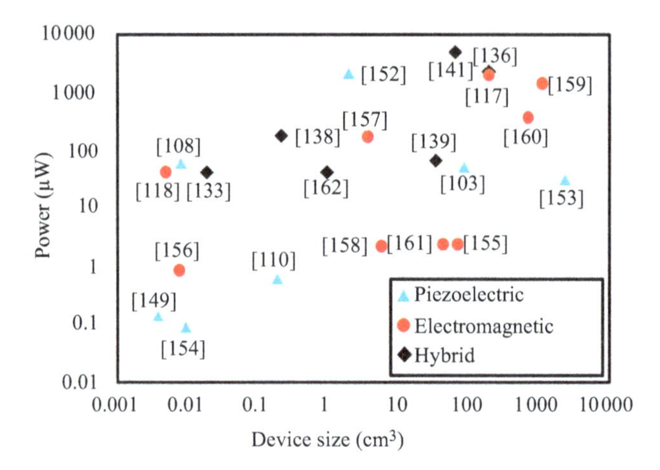


*Figure 3.31 Power versus device size comparison of the reported EHs*

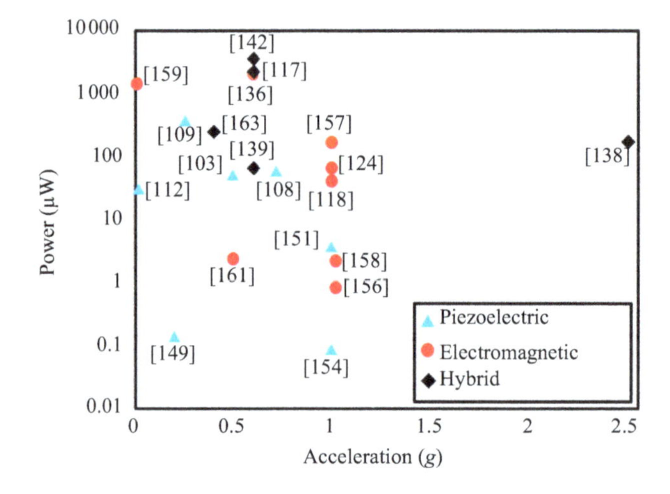


*Figure 3.32 Power produced by the reported EHs as a function of the acceleration levels*

because of their dual transduction mechanisms. Also worth highlighting is that the harvesters reported in [135,179] produce high power despite their low base accelerations due to their sizes, as obtained from the previous figure. The harvesters reported in [122,128,130,157,179,181,184] are low-acceleration harvesters (less than 0.5 g), while the harvesters reported in [127,136,142,170,173,176–178] are designed and characterized at medium acceleration levels (from 0.5 to 1 g). Only harvester reported in [156] can be categorized into a relatively high acceleration device since it was tested at an acceleration level greater than 2 g.

Figure 3.33 illustrates power versus resonant frequency for the developed EHs. Environmental vibrations are usually random, characterized by low frequencies and

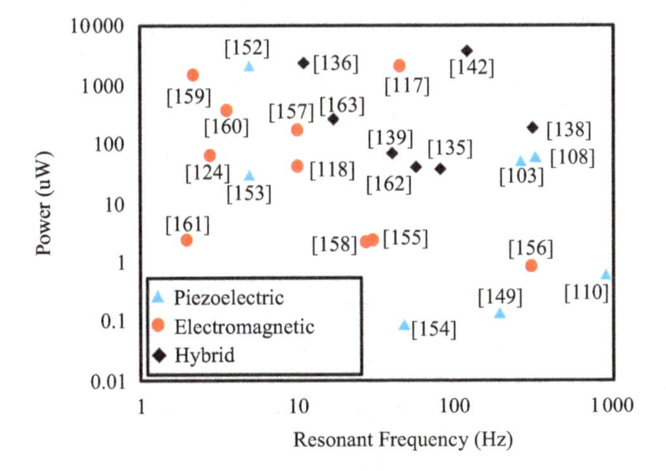


*Figure 3.33 Harvested power as a function of the harvester's resonant frequency*

low amplitudes, which is the reason for the EHs need to be designed to resonate at lower frequencies so that it will perform better in practical applications. EH generates peak power when its natural frequency matches the frequency of the vibrating structure. The hybrid transduction mechanism, however, adds to the increased power output as compared to the stand-alone PEEH and EMEH harvesting, as these harvesters operate at wide frequency bandwidths. The highest power shown in the graph, in the case of [154], is attributed to hybrid conversion at low-frequency oscillations. Most of the EMEHs resonate at lower frequencies as depicted in the graph. However, PE generators in [122,127] resonate at high frequencies and produce respectable power except for EHs in [16,168,173], due to their comparatively minute dimensions.

A comparison based on power as a function of the internal impedance of the reported EHs is shown in Figure 3.34. With an increase in internal impedance, the output harvested power decreases as the less current level is available at the harvester's output terminals. The power produced by EMEHs [135,179,180] is high since these EMEHs have comparatively less internal resistances. PEEHs, generally, have relatively higher internal impedances, which consequently reduce the power generations. In most of the reported HEHs, such as in [151,154,156–158,159,161,183,184], the PE and triboelectric components contribute to their high internal impedances. However, as their hybrid nature means these devices have more power-harvesting capabilities, output power generation from these EHs is generally high.

Figure 3.35 demonstrates the relationship between power density and base acceleration of the reported EHs, which are characterized under a range of base accelerations from 0.01 to 2.5 g. The power density increases with an increase in generated power but decreases with an increase in the harvester's dimensions. The PEEHs developed in [127,165] produces high power despite their smallest sizes, which consequently gives the highest power density among the reported EHs. At low base accelerations of 0.1–0.5 g, the harvesters investigated in [139,168,181,183] exhibit higher power densities than most of the reported devices because of their higher power

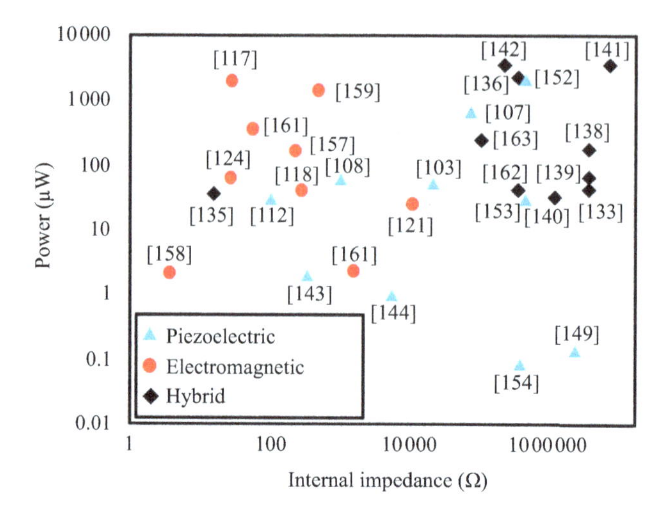


*Figure 3.34 Reported EHs power versus harvester's internal impedance*

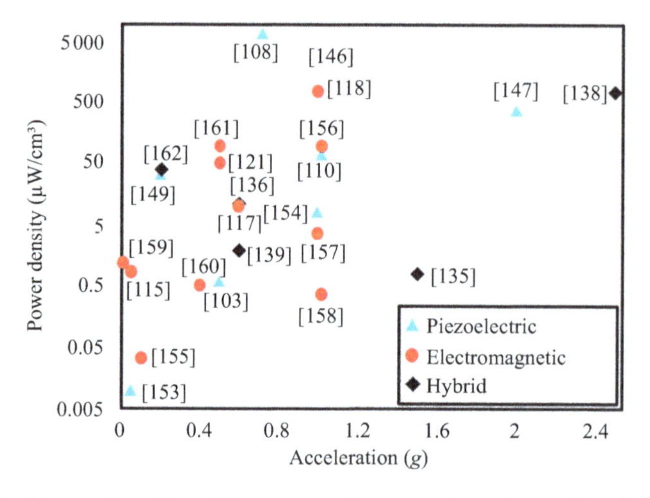


*Figure 3.35 Power densities with respect to base accelerations of the developed EHs*

generation levels with comparatively small sizes. The EMEHs presented in [136,176], the PEEHs reported in [16,166], and the hybrid EHs reported in [156] have more power densities due to their small dimensions, while the EHs reported in [172,175,178] have lower power densities because of their larger comparative sizes.

The power density per acceleration values of the harvesters reported in the literature are examined with respect to their internal impedances, as shown in Figure 3.36. The EMEHs reported in [136,139,179,181] and the PEEHs reported in [127,166,168], and a few of the hybrid harvesters reported in [156,183] have a high power density per acceleration values than the PEEHs reported in [122,172,173], the EMEHs reported in [135,133,177,178,179], and the HEHs reported in [153,154,157]. Generally, the internal impedances of PEEHs are more than those of the EMEHs as represented in the graph.

The power density per acceleration values of the harvesters are also compared with their resonant frequencies, to study the overall trend of the PEs, EMs, and HEHs. These values range from 0.2 $\mu$W/g·cm$^3$ to the highest value 9 697.55 $\mu$W/g·cm$^3$, with PEEH reported in [127] exhibiting the best performance. Generally, the EMEHs operate on lower resonant frequencies than both hybrid and PEEHs cited in the literature, except for the PEEH reported in [172], as shown in Figure 3.37.

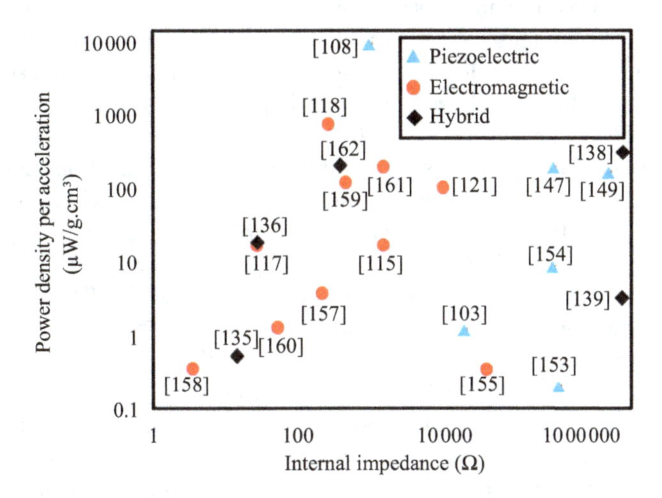


*Figure 3.36 Power density per acceleration versus internal impedance of the reported EHs*

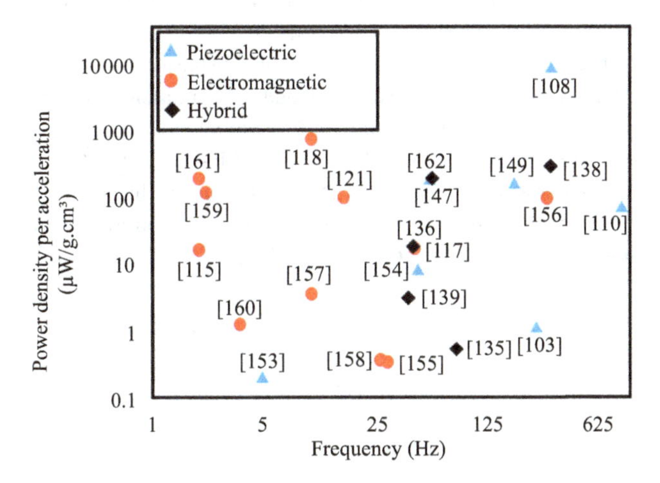


*Figure 3.37 Power density per acceleration as a function of resonant frequencies of the reported EHs*

## 3.4 Summary

Traditional Li-ion batteries are incapable of coping with the power requirements of the excessively used smart electronic devices and the increased usage of WSNs. To solve this problem, researchers have been looking into energy harvesting as a means to provide power, with emotion-driven devices representing a large fraction of this sustainable solution. Material selection, operational frequency bandwidth, beam deflection amplitude, voltage, and load resistance values are important characteristics of a VEEH that need to be considered, as it affects the output power of the EHs. In terms of material selection for the PEEHs, the commonly used PZT offers several advantages over other PE materials. Other than its higher power density and low hysteresis losses, it also offers a good signal-to-noise ratio. The introduction of nonlinear PE coupling can improve the system performance and increase the power-generation capability. Most of the reported harvesters in the literature are nonlinear and resonant type. However, some nonresonant VEHs which perform better in off-resonant states and eliminate the use of proof mass are also discussed. EMEHs generally comprised coil and magnet arrangement and produce an AC voltage as a result of the changing magnetic flux across the coil when there is a relative motion between the magnet and coil. These systems performed well in low-frequency resonant-based mechanical vibrations, presented with good electromechanical coupling, and are capable of producing more power than PEEHs due to low internal impedance of the coil. Considerable works have also been performed on combining more than one harvesting mechanism, by harvesting simultaneous energy from the same source or different sources to tune up the resonant frequencies of microharvesters. This hybrid technology results in increased operating frequency bandwidths and improved power densities. With an increasing research interest in biomechanical energy harvesting, outstanding performance and successful integration of emotion-driven hybrid EHs into wearable electronics is currently undertaken by the research community and may be improved with the introduction of a spiral spring instead of a cantilever beam as a suspended element to achieve the lowest resonant frequencies of human vibrations. Harvester's assembly may be modified by increasing the magnet's size as well as the number of coils' turns in EMEHs. Hybrid, PE-EM, PE-TE, and PE-EM-TE conversion mechanisms may be combined in the single harvester to improve its performance and applicability. The power-conditioning circuit needs to be incorporated into the existing systems to ensure efficient energy conversion.

# Chapter 4
# Design and modeling of vibration energy harvesters

## 4.1 Introduction

Microsensors and portable electronics are more ubiquitous in the current mobile world, which increased the demand for long-lasting and self-sustained power supply with reduced overall system's size. To address this demand, energy harvesting from vibrating bodies has attracted significant attention from both the academia and industry to make these wireless transceivers and wearable gadgets autonomous for their continuous and uninterrupted health-monitoring applications. Machines, environmental structures, and humans have abundant sources of energies in the form of potential, thermal, mechanical, and chemical energies, which can be easily harvested by novel miniaturized fabrication techniques. Piezoelectric, electromagnetic, electrostatic, triboelectric, and hybrid energy harvesting are the most promising and reliable techniques among others [185]. A notable fraction of the wearable electronic devices used for health monitoring, diagnostics, and diseases prevention are still dependent on batteries [186]. The batteries, due to their bounded lifespan, network congestion, maintenance cost, and large volume, limit the use of these smart devices, especially in embedded and remote locations, and are not ideal anymore.

VEHs are typically comprised of a mass–spring arrangement or a cantilever beam with an attached proof mass. The oscillating spring or cantilever beam on exposure to ambient vibrations allows the conversion of mechanical energy into electrical energy by either electromagnetic induction due to the relative motion of the coil and magnet, piezoelectricity by the direct piezoelectric effect, which is then transferred to an externally connected circuit or a hybrid piezo-electromagnetic dual transduction [187].

VEHs are considered to be a promising solution for the long-lasting operation of WSNs, moreover, making these power-sustainable in hazardous, implantable, remote, harsh, embedded, and abandoned environments where frequent battery replacement is a challenge. With the advancement in microelectronics, microsensors, and low-power wireless transmission technology, WSNs are broadly used in many fields such as biomedical devices, environmental monitoring, military assets, personal tracking, structural health monitoring of bridges and civil infrastructure, engine monitoring of aircrafts, machines, and recovery systems [188]. The WSNs consume electrical power for the operation of onboard microsensors, signal-processing circuits, power management circuits, memory, microcontrollers, and transceiver.

A promising alternative to batteries is energy harvesting from the ambient environment of the WSNs by the use of energy harvesters. Energy can be harvested by electromechanical transduction mechanisms, such as, electrostatic, electromagnetic, or PEH. PEH is relatively the simplest and efficient among the other transduction techniques. Piezoelectric elastic strips are usually encapsulated in metallic plates acting as electrodes for the harvester. Piezoelectric plates produce electrical energy by its property of deformation, when pressure, force, or mechanical strain is applied on it. A low-frequency resonating piezoelectric bender is best suited in design consideration when harvesting energy from the body kinetics of low frequency (<25 Hz) and high amplitude [189].

A PEEH reported by Renaud *et al.* [190] is comprised of a flexible piezoelectric strip. The harvester generate voltage across the bendable PZT strip when subjected to foot strike. Two piezoelectric capacitors encapsulate an elastic cantilever beam. The prototype has 60 g weight, 14 $cm^3$ volume and when subjected to a base excitation of 10 cm and 10 Hz of moving hand, it produced an output power of 600 μW and 47 μW power by rotating at 180°/s.

An electromagnetic-type energy generator that harvests energy from human body kinetics is developed by Wang *et al.* [138]. A 25 mm hollow tube enclosing a permanent magnet (20 mm diameter and 4 mm thick) and two similar magnets were attached to opposite ends of the tube. It is fabricated from Teflon using 3D printing, and a circular coil having 480 turns is wrapped around the tube. When the prototype is subjected to vibrations, the magnets start movement due to the repulsive force of the middle cylindrical magnet and an induced EMF is generated at the coil terminals according to Faraday's law of electromagnetic induction. The tube with the coil weighs 103.7 g. When 5 Ω resistance which is equal to the coil internal resistance was connected across it and characterized at a base excitation of 0.35 g, a voltage of 0.34 V was generated. A maximum voltage of 0.86 V is delivered to a matching load at 9.155 Hz and under 0.85 g base acceleration level.

The harvester is connected vertically and then horizontally to a test subjects' right leg which produced a voltage level of 0.2245 V and 0.42 mW power at a walking speed of 5 km/h, while 0.46 V and 6.1 mW at a running speed of 9 km/h, when the leg strikes the ground. A maximum output power of 10.6 mW is produced by the harvester when it is attached transversely to the leg and at 8 km/h it is expected to be increased with increasing subject speed and the number of coil turns.

A body-worn sensor network for health monitoring is empowered by von Büren [189], which produced an electrical voltage by the electromagnetic coupling of a translator and a stator across the coil terminals. The translator also acts as a proof mass to initiate movement for the EMEH, and a bearing is used to guide its motion. The bearing separates it from the stator and acts as an elastic spring for the returning motion of the translator. This bearing is attached to the translator by two rods, and the coil is connected to a PCB circuit. Screws are used to adjust the gap of the coil to the generator housing easily, and the total structure weighs 43 g and occupies a volume of 0.49 $cm^3$.

The prototype was characterized inside the lab with a vibration shaker for sinusoidal motion, and then under initial base excitations and finally worn by a test subject (human). The internal resistance of the device is measured as 5 kΩ, and when 10 kΩ is connected across the coil terminals, it produced 5.1 μW power and 22.0 μW, when mounted on the upper arm and below the knee joint, respectively, during normal walk, which is sufficient to power up most of the body-worn sensors. The device produced 0.5–4 V load voltage levels and 5–25 μW power.

A piezoelectric-type human motion-driven energy harvester is produced by Wei *et al.* [122], using a high-frequency lead zirconate titanate (PZT-5A) bimorph cantilever beam with a proof mass as deflection force at the free end. A ridged cylinder encloses a shaft, and the PZT-5A with a tip mass is attached to one end of the shaft. When the cylinder is excited by an impulse movement force, the PZT-5A bimorph starts oscillating continuously due to striking with multiple ridges, and a resultant electrical output can be obtained from its terminals. The harvester is encapsulated in an aluminum rectangular cove, and rubber is used for the fragile PZT to be safe at the time of excitations; the overall dimension of the prototype is 90 mm × 32 mm × 24 mm and the prototype weighs 234 g. The resonant frequency of the bimorph is 260 Hz, and the ossciloscope is set at 10 times higher sampling rate. At a normal walking speed of 5 km/h on a treadmill, and when 20 kΩ resistance is connected, which is equal to the PZT optimum resistance, a maximum power of 51 μW is generated.

Piezoelectric, electromagnetic, and electrostatic energy-harvesting techniques can be used for insole energy harvesting, but due to the thickness of device's architecture of the latter two techniques, PEEHs are best suited to be integrated into the shoe [191].

## 4.2 Design and modeling

### *4.2.1 Architecture and the working mechanism*

A vibration energy harvester can be modeled as a transverse beam with a magnet and an airfoil as the tip mass, which is shown in Figure 4.1.

Undamped natural frequency

$$\omega_n = \sqrt{\frac{k}{m}} \tag{4.1}$$

is derived in terms of the proof mass $M$ (total mass of the magnet and airfoil) and the stiffness

$$k = \frac{3\,EI}{L^3} \tag{4.2}$$

of the beam. In the above equation, $E$ represents the Young's modulus of elasticity, where $E = 200$ GPa for hot dipped galvanized (GI) steel, and $L$ is the

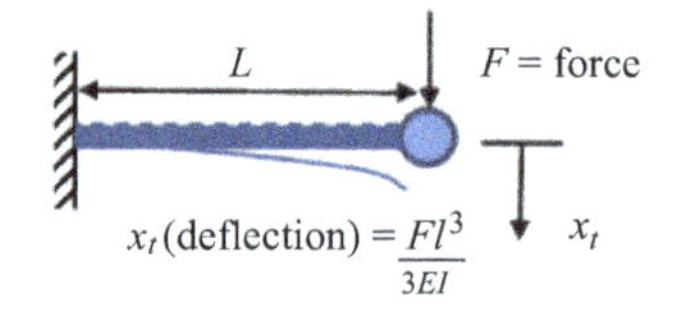


(a)

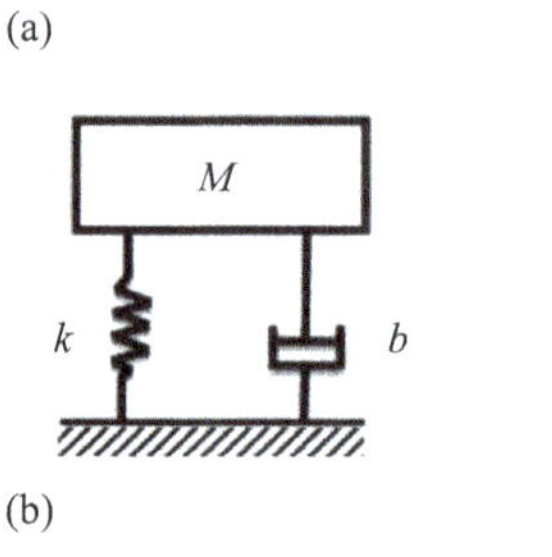


(b)

*Figure 4.1 Model of the VEH: (a) simplified model and (b) an equivalent mass–spring–damper system*

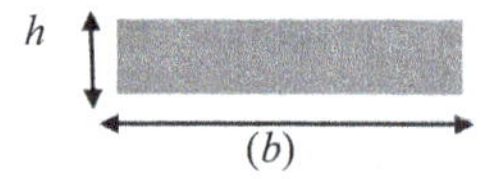


*Figure 4.2 Cross-sectional view of the cantilever beam for VEHs*

length of the cantilever and

$$I = \frac{bh^3}{12} \tag{4.3}$$

is the mass moment of inertia. $b$ is the length and $h$ is the height of beam cross-section (Figure 4.2).

$$k = \frac{3\,Ebh^3}{12\,L^3} \tag{4.4}$$

$$k = \frac{Ebh^3}{4\,L^3} \tag{4.5}$$

$$\omega_n = \sqrt{\frac{Ebh^3}{4\,mL^3}} \tag{4.6}$$

The mass flow rate is

$$\dot{m} = \rho\dot{Q} \tag{4.7}$$

$$\dot{m} = \rho a V_{\text{air}} \tag{4.8}$$

$\rho$ is the density of air

$$\rho = 1.225\,\frac{\text{kg}}{\text{m}^3} \tag{4.9}$$

and $a$, the area of flow is given by

$$a = \frac{\pi}{4} d_{\text{pipe}}^2 \tag{4.10}$$

$$a = 0.01266\ \text{m}^2 \tag{4.11}$$

$$\dot{m} = 1.225 \times 0.01266 \times 9 \tag{4.12}$$

$$\dot{m} = 0.13\ \text{kg/s} \tag{4.13}$$

$$k_{\text{eq}} = \frac{3EI}{L^3} \tag{4.14}$$

$$m_{\text{eq}} = \frac{33}{140}(m_L + M_t) = 0.09\ \text{kg} \tag{4.15}$$

$$\omega_n = \sqrt{\frac{k}{m}} \tag{4.16}$$

$$f_n = \frac{1}{2\pi}\omega_n \tag{4.17}$$

$$f_n = \frac{1}{2\pi}\sqrt{\frac{k}{m}} \tag{4.18}$$

Bridges vibrate in a frequency range of 1–40 Hz, hence

$$f_n = 1 - 40\ \text{Hz} \tag{4.19}$$

$$k = (2\pi f_n)^2 \times m \tag{4.20}$$

$$\frac{3EI}{L^3} = m\,(2\pi f_n)^2 \tag{4.21}$$

$$\frac{3Ebh^3}{12L^3} = m\,(2\pi f_n)^2 \tag{4.22}$$

$$L^3 = \frac{3\ Ebh^3}{12\ m\,(2\pi f_n)^2} \tag{4.23}$$

$$L = \left(\frac{3\ Ebh^3}{12\ m(2\pi f_n)^2}\right)^{1/3} \tag{4.24}$$

$$h = 0.0005\ \text{m}, b = 0.025\ \text{m} \tag{4.25}$$

$$L = 140\ \text{mm (for a resonant frequency of 3.6 Hz)} \tag{4.26}$$

Length of beam-II is 5.5 cm, which corresponds to a resonant frequency of about 45 Hz.

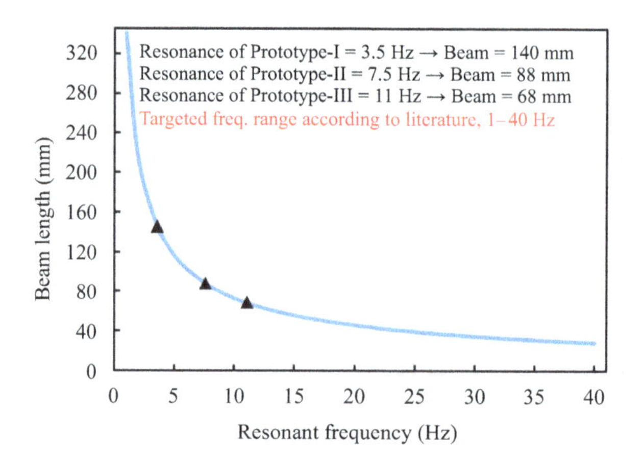


*Figure 4.3 Resonant frequency of the harvester as a function of the beam length*

Figure 4.3 shows the simulated resonant frequency of the harvester versus length of the cantilever beam. The length of cantilever beam can be modeled for device's resonant frequencies. Knowing the resonant frequencies of the energy harvester, the dimension of the cantilever beam used can be determined using (4.24). The cantilever modeling is validated by simulation results. At an input frequency (resonance, 3.6 Hz) of the first prototype, the length of the beam is at 140 mm, which is exactly equal to the length of the fabricated beam. Similarly, this model analysis can be applied to the second and third prototypes, which also satisfy the simulation results of the analytical modeling of (4.24).

Bending stresses

$$\sigma_b = \frac{M}{Z} \tag{4.27}$$

$$Z = \frac{I}{C} \tag{4.28}$$

where $C$ is the distance from the neutral axis.

$$\sigma_b = \frac{M_c}{I} \tag{4.29}$$

$$\sigma_{b(\max)} = \frac{M_{\max} b/2}{wb^3/12}; c = \pm\frac{b}{2} \tag{4.30}$$

$$\sigma_{b(\max)} = M_{\max}\frac{b}{2} \times \frac{12}{wb^3} \tag{4.31}$$

$$\sigma_{b(\max)} = \frac{M_{(\max)}(6)}{wb^2} \tag{4.32}$$

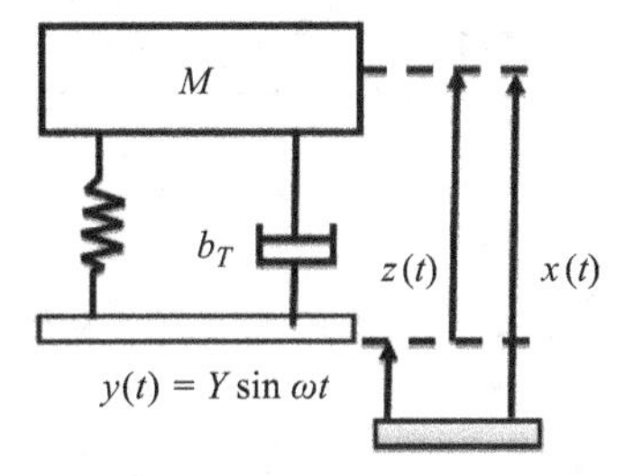


*Figure 4.4 Lumped parameter model of the VEH*

The cantilever beam design equation, with tip mass, as force

$$\sigma_{b(\max)} = \frac{6FL}{wb^2}; \ M_{\max} = FL \tag{4.33}$$

On subjecting the harvester to base excitation as shown in Figure 4.4, the displacement of the magnet (cantilever beam) is given by

$$X = \left( \frac{Y\omega^2}{\omega^2{}_n \sqrt{\left(1 - \left(\frac{\omega}{\omega n}\right)^2\right)^2 + \left(2\xi_T \frac{\omega}{\omega n}\right)^2}} \right) \tag{4.34}$$

is derived in terms of the excitation frequency $\omega$, the amplitude of base excitations $Y$, the damping ratio $\xi_T$, and the resonant frequency $\omega_n$ of the harvester.

$$G = \left( \frac{A\omega}{\omega^2{}_n \sqrt{\left(1 - \left(\frac{\omega}{\omega n}\right)^2\right)^2 + \left(2\,\xi_T \frac{\omega}{\omega n}\right)^2}} \right) \tag{4.35}$$

By Faraday's law of electromagnetic induction [192], the generated voltage

$$V(\omega) = -G \frac{dB_x}{dx} S \tag{4.36}$$

as the frequency function can be calculated for the harvester in terms of the relative velocity of the magnet $G$, the magnetic flux density $B_x$, and the area of coil turns $S$.

The magnetic flux density $B_x$, along the normal line to the center of a cylindrical magnet, is [193]

$$B_x = \frac{B_r}{2} \left( \frac{x + H_m}{\sqrt{(x + H_m)^2 + (r_m)^2}} - \frac{x}{\sqrt{(x^2 + r_m{}^2)}} \right) \tag{4.37}$$

depends on the height of the magnet $H_m$, flux density $B_r$, radius $r_m$ of the magnet, and the distance $x$ from the magnet which can be modified as

$$\frac{dB_x}{dx} = \frac{B_r}{2}\left(\left(\frac{D_1 + H_m}{\sqrt{(D_1 + H_m)^2 + (r_m)^2}} - \frac{(D_1 + H_m)^2}{\sqrt[3]{(D_1 + H_m)^2 + (r_m)^2}}\right) - \left(\frac{1}{\sqrt{{D_1}^2 + {r_m}^2}} - \frac{{D_1}^2}{\sqrt[3]{{D_1}^2 + {r_m}^2}}\right)\right) \tag{4.38}$$

for a wound coil, with $N$ number of turns, inner radius $r_p$ and diameter $d_w$ of the wire, and area $S$ sum can be calculated as:

$$S = \sum_{i=1}^{N} S_i \approx \sum_{i=1}^{N} \pi {r_i}^2 \tag{4.39}$$

$$r_i = r_p + \left(i - \frac{1}{2}\right) d_w \tag{4.40}$$

obtained by taking the derivative for $x$ and substituting its value in $D_1$ (gap between the magnet and the coil).

For a multilayered wound coil as shown in Figure 4.5, the time response of voltage gain can be converted to frequency domain as

$$V(\omega) = -G\sum_{i=0}^{n} \frac{B_r}{2}\left(\left(\frac{D_i + H_m}{\sqrt{(D_i + H_m)^2 + (r_m)^2}} - \frac{(D_i + H_m)^2}{\sqrt[3]{(D_i + H_m)^2 + (r_m)^2}}\right) - \left(\frac{1}{\sqrt{D_i^2 + {r_m}^2}} - \frac{{D_0}^2}{\sqrt[3]{D_i^2 + {r_m}^2}}\right)\right) S \tag{4.41}$$

as a function of $D_i$, distance of the single layer from the magnet.

Peak voltage on the connecting load is

$$V_{L\text{peak}} = \left(\frac{R_L}{R_L + R_C}\right) V_{\text{peak}} \tag{4.42}$$

and average load power $P_L$ across the load resistance $R_L$

$$P_L = \frac{{V_L}^2}{2R_L} \tag{4.43}$$

to the load depends on, the load impedance $R_L$ and coil impedance $R_C$.

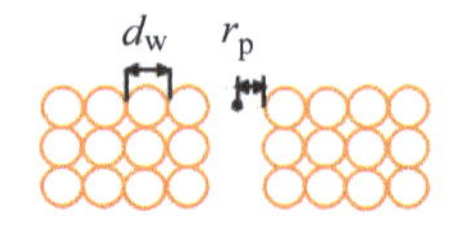


*Figure 4.5 Multilayered wound coil cross-section*

### *4.2.2 Finite element modeling*

The voltage produced from the electromagnetic portion of the harvester depends on the residual flux density, coil turns, and the overall coil turn areas. In order to accurately design and place the coil in the harvester, the permanent magnet is simulated in Vizimag software. In Figure 4.6, the magnetic flux density distribution along the magnets cross-sectional plane is depicted. It is clear from Figure 4.6 that the magnetic flux density is mostly concentrated near the surface of the magnet. However, away from the magnet the flux density decreases significantly.

In Figure 4.6, the magnetic flux density as a function of distance from the magnetic surface along its axis is shown. The simulation predicts that the magnetic flux density (about 6 000 Gauss) is maximum near the surface of the magnet. However, away from the magnet's surface, the flux density drastically decreases to zero Gauss.

Since the magnetic flux density gradient, as shown in Figures 4.6 and 4.7, is relatively high near the magnet's surface, the coil should be placed nearer to the magnet in order to achieve a high output voltage. However, in order to allow the oscillations of the magnet over the coil, the gap should be adjusted according to the amplitude of vibration of the magnet at resonance. The distribution of the normal component of magnetic flux density at different distances (5, 10, and 15 mm) from the magnet is shown in Figure 4.8. In the figure, the horizontal distance of about 19 mm (from −9.5 to 9.5 mm) corresponds to the diameter of the permanent magnet. It is obvious from the figure that the magnetic flux density within the diameter of the magnet (from −9.5 to 9.5 mm) is maximum at all the three gaps (5, 10, and 15 mm). However, on both sides of the magnet the magnetic flux density decreases rapidly. The relatively low magnetic flux density (on the sides of the magnet) within the area of the coil during operation of the harvester will lead to reduced voltage production in the outer turns of the coil. In addition, the power production from the harvester will also decrease as the outer turns of coil contribute to the overall impedance of the coil and results in high power loss in the coil. By accounting for the flexibility of the cantilever beam and keeping the high magnetic flux density over the entire coil turns, the gap between the coil and the magnet is kept at 10 mm, and the coil diameter is selected as 20 mm.

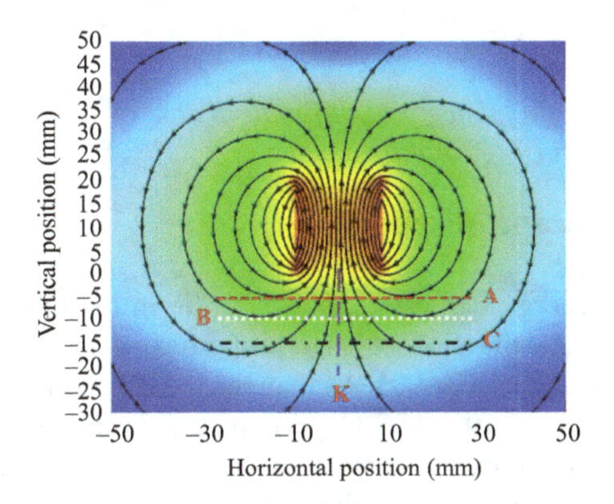


*Figure 4.6 Magnetic flux density distribution*

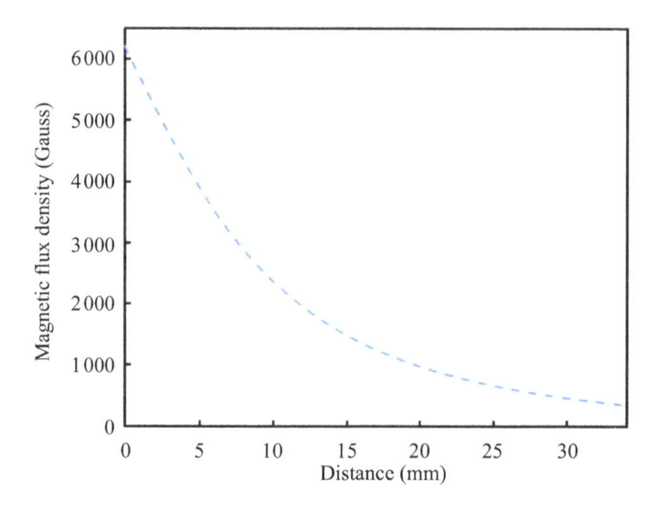


*Figure 4.7 Variation in magnetic flux density as a function of distance from the magnet's surface (line K in Figure 4.6)*

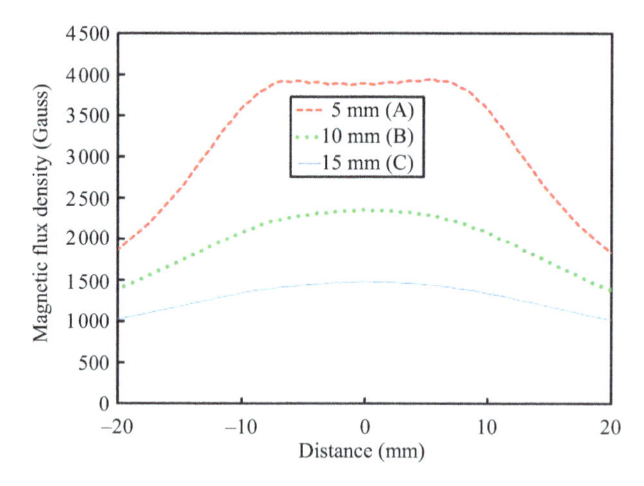


*Figure 4.8 Magnetic flux density at different distances from the magnet (lines A, B, and C in Figure 4.6)*

## 4.3 Comparison and discussion

Piezoelectric insole energy harvesters, however, lie in the higher frequency range (5.6 Hz–20 kHz) due to the piezoelectric properties of these harvesters. EMEHs produced more power (61.3–10 400 μW) than that of PEEHs (43–51 μW) due to the lower internal impedance (10–104.7 Ω) in comparison to piezoelectric generators with internal impedance ranging from 150 Ω to 20 kΩ. Table 4.1 surveys a comparison of various materials along with their key characteristics and performances for different energy harvesters selected from the relevant literature.

*Table 4.1 Comparison of materials, characteristics and performances for different energy harvesters selected from the relevant literature*

| Material | Volume ($cm^3$) | Resonant frequency (Hz) | Base acceleration (g) | Internal resistance (Ω) | Voltage (mV) | Power (μW) | Power density ($\mu W/cm^3$) | Ref. |
|---|---|---|---|---|---|---|---|---|
| PZT-5H | — | 542 | 1.12 | 100k | — | — | — | [118] |
| PZT ($d_{11}$) | | | | | | | | |
| PZT ($d_{33}$) | | 225.9 | 2.5 | 150 | 1.792 | 2.765 | | |
| PZT-5A | 1.06, 1.52 | 85–100 | — | 300, 225 | 2.36, 2.25 | 18.5, 22.5 | 17.45, 14.80 | [129] |
| Prototype-I | | | | | | | | |
| Prototype-II | | | | | | | | |
| PZT-5H and Cu | — | 21, 31 | 0.1 | 60k, 29k | 5, 3 | 700, 530 | — | [194] |
| PEH-1 | | | | | | | | |
| PEH-2 | | | | | | | | |
| PZT | | | | | | | | |
| Linear-1 | 1.321 | 28 | 0.36 | 10k | 545.244 | 29.729 | 22.50 | [195] |
| Linear-2 | | 80.4 | | | 484.963 | 23.519 | 17.80 | |
| Nonlinear | | — | | | 641.041 | 41.093 | 31.10 | |

## 4.4 Summary

Evolutional advances in wireless sensor network technology have been driven by issues raising from battery dependency in microelectronic systems. Energy harvesting is a growing research interest to improve the lifespan of batteries or to reduce its use, and motion-driven devices represent a large contributing factor. Piezoelectric materials absorb the available mechanical energies, usually vibrations in the surrounding to convert it into electrical energy for powering low-energy operated devices. These harvesting devices are fit to be kept in far off and remote locations, such as structural monitoring sensors and GPS tracking devices on animals in the wild. The introduction of nonlinear piezoelectric coupling also adds improvement to the system's performance and better output power. A considerable work has also been reported on using different approaches to tune the resonant frequencies of microgenerators using FUC for increasing the frequency range of these harvesters. Harvesting energy from human motion can boost up the performance and reliability of wearable devices by making them self-sustainable. The ultrasonic-based wireless transmission energy harvesting is relatively safe to humans and by this approach energy can be extracted in two dimensions, i.e., the $x$ and $y$-axes, which doubles the harvester's efficiency.

# Chapter 5
# Nonlinear 3D printed electromagnetic vibration energy harvesters

## 5.1 Introduction

Microsensors and portable electronics are more ubiquitous in the current mobile world and resulted in increased demand for long-lasting and self-sustained power supply with reduced overall system size. To address this demand, energy harvesting from living organisms has attracted significant attention from both academia and industry to make these wearable devices and wireless transceivers autonomous for their continuous and uninterrupted health-monitoring applications. Animals and humans have abundant source of energies in the form of, thermal, mechanical, and chemical energies, which can be easily harvested using novel miniaturized fabrication techniques. Foot-induced excitations are particularly attractive among the other body movements for scavenging considerable energy.

The power consumption of different sensors is listed in Table 5.1.

Piezoelectric, electromagnetic, and electrostatic energy harvesting are the most promising and reliable techniques among others [180]. A notable fraction of the wearable electronic devices used for health monitoring, diagnostics, and disease prevention are still dependent on batteries [186]. The batteries, due to their bounded lifespan, network congestion, maintenance cost, and large volume, limit the use of these smart devices especially, in embedded and remote locations and are not ideal anymore. A promising alternative to batteries is energy harvesting from the ambient environment of the wireless sensors nodes by the use of EHs [3].

Energy can be harvested by electromechanical transduction mechanisms, such as electrostatic, electromagnetic, or PEH. PEH is relatively the simplest and efficient harvesting technique [96,196–198]. In PEHs, piezoelectric elastic strips are usually encapsulated in metallic plates acting as electrodes and produce electrical

*Table 5.1 Power consumption of different sensors [154]*

| Sensor type | Minimum voltage (V) | Minimum current ($\mu$A) | Power ($\mu$W) |
|---|---|---|---|
| Pressure | 1.8–2.1 | 1–4 | 1.8–8.4 |
| Temperature | 2.1–2.7 | 0.9–14 | 1.89–37.8 |
| Acceleration | 1.8–2.5 | 10–180 | 21.6–324 |

energy by its property of deformation as a result of an applied pressure, force, or mechanical strain [199]. A low-frequency resonating piezoelectric bender is best suited in design consideration when harvesting energy from body kinetics of low-frequency ($<$25 Hz) and high-amplitude vibrations.

A PEEH is designed and fabricated by Renaud *et al.*, which is based on a bending piezoelectric strip. Two piezoelectric capacitors encapsulate an elastic cantilever beam, which is reported in [190]. The prototype has 60 g weight, 14 $cm^3$ volume and when subjected to a base excitation of 10 cm and 10 Hz of moving hand, an output power of 600 $\mu$W was produced. The harvester can produce electrical power by limb movement and when rotated at 1 800 per second, it produced 47 $\mu$W power.

An electromagnetic type body-worn kinetics energy generator is developed by Wang *et al.* [138]. A 25 mm hollow tube enclosing a permanent magnet (20 mm diameter and 4 mm thick) and two similar magnets were attached to opposite ends of the tube. It is fabricated from Teflon using 3D printing, and a circular coil having 480 turns is wrapped around the tube. When the prototype was subjected to vibrations, the magnets start moving due to the repulsive force of the middle cylindrical magnet and an induced EMF is generated at the coil terminals according to Faraday's law of electromagnetic induction. The tube with the coil weighs 103.7 g. When a resistance of 5 $\Omega$, equal to the coil internal resistance, was connected across it and characterized at a base excitation of 0.35 g, 0.34 V was produced. A maximum voltage of 0.86 V was delivered to a matching load at 9.155 Hz and under 0.85 g base acceleration level.

The harvester presented in [189] was connected vertically and then horizontally to a test subjects' right leg and produced a voltage level of 0.2245 V and 0.42 mW power at a walking speed of 5 km/h, while 0.46 V and 6.1 mW at a running speed of 9 km/h when the leg strikes the ground. A maximum output power of 10.6 mW was produced by the harvester at 8 km/h when attached transversely to the leg and the output power is expected to be increased with increasing subject's speed and the number of coil turns.

A body-worn sensor network reported in this research produced electrical voltage by the electromagnetic coupling of a translator and a stator across the coil terminals. The translator acts as a proof mass to initiate movement for the EMEH, and a bearing is used to guide its motion. The bearing separates it from the stator and acts as an elastic spring for the returning motion of the translator. This bearing is attached to the translator by two rods, and the coil is connected to a PCB circuit. Screws were used to adjust the gap of the coil to the generator housing easily, and the total structure weighs 43 g with a volume of 0.49 $cm^3$.

The prototype was characterized inside the lab with a vibration shaker for sinusoidal motion, then under initial base excitations and finally worn by a test subject (human). The internal resistance of the device is measured as 5 k$\Omega$, and when 10 k$\Omega$ was connected across the coil terminals, it produced 5.1 $\mu$W and 22.0 $\mu$W power, when mounted on the upper arm and below the knee joint, respectively, during the normal walk, which is sufficient to power up most of the body-worn sensors. The device produced 0.5–4 V load voltage levels and 5–25 $\mu$W power.

A piezoelectric-type human motion-driven EH is produced by Wei *et al.* [122], using a high-frequency lead zirconate titanate (PZT-5A) bimorph cantilever beam with a proof mass as deflection force at the free end. A ridged cylinder enclosed a shaft, and the PZT-5A with a tip mass was attached to one end of the shaft. When the cylinder is excited by an impulse movement force, the PZT-5A bimorph starts oscillating continuously due to striking with multiple ridges, and a resultant electrical output was obtained at its terminals. The harvester was encapsulated in an aluminum rectangular cove, and rubber is used for the fragile PZT to be safe at the time of excitations. The overall dimension of the prototype is found to be 90 mm × 32 mm × 24 mm and the prototype weighs 234 g. The resonant frequency of the bimorph recorded as 260 Hz, and the oscilloscope was set at 10 times higher sampling rate. At a normal walking speed of 5 km/h on a treadmill and when 20 kΩ (which is equal to the PZT optimum resistance) was connected, it generated a maximum power of 51 $\mu$W.

Piezoelectric, electromagnetic, and electrostatic energy-harvesting techniques can be used for insole energy harvesting, but due to the thickness of the device's architecture in EM and ESEHs, PEEHs are best suited to be integrated into the shoe [191].

## 5.2 Design and modeling

### *5.2.1 Architecture and the working mechanism*

A cross-sectional view of the proposed EH is shown in Figure 5.1. It is a MDOF system, and each beam can be modeled as a SDOF system for simplicity of equations.

With foot strike, the wound coil and magnets attached to the cantilever beam start oscillating and an induced emf will be generated across the coil.

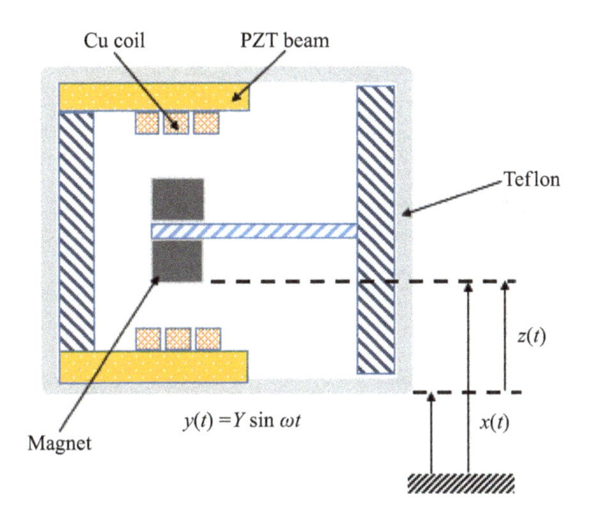


*Figure 5.1 A cross-sectional view of the proposed EH*

GI steel was considered for the fabrication of cantilever beam with Young's modulus of elasticity, $E = 200$ GPa. The natural frequency of the beam can be determined from the beam's length equation with width $b$ and thickness $h$.

The beam length $L$ is determined based on:

$$L = \left( \frac{3Ebh^3}{12M(2\pi f_n)^2} \right)^{1/3} \tag{5.1}$$

where $M$ is the total mass of the beam with tip mass and $f_n$ is the natural frequency of oscillations. Equation (5.1) can be simulated to give beam length as a function of natural frequency and a linear relationship between the frequency and beam length is depicted in Figure 5.2. For fundamental frequencies from 1 to 12 Hz, the beam length can be altered from 340 to 73 mm, as shown in Figure 5.2.

The beam length can be related to the resonant frequency based on the excitation levels to which the harvester is subjected to.

The cantilever beam can also be modeled as SDOF with base acceleration amplitude $A$ [200] between the coil and the magnet, which is dependent on $\omega$ (base excitation frequency), $\omega_n$ (undamped natural frequency), and $\omega^2{}_n\xi_T$ (total damping ratio).

$$G = \left( \frac{A\left(\frac{\omega}{\omega n}\right)^2}{\omega^2{}_n\sqrt{\left(1-\left(\frac{\omega}{\omega n}\right)^2\right)^2 + \left(2\xi_T\frac{\omega}{\omega n}\right)^2}} \right) \tag{5.2}$$

At resonance when $\frac{\omega}{\omega n} = 1$, then

$$G = \left( \frac{A}{2\omega^2{}_n\xi_T} \right) \tag{5.3}$$

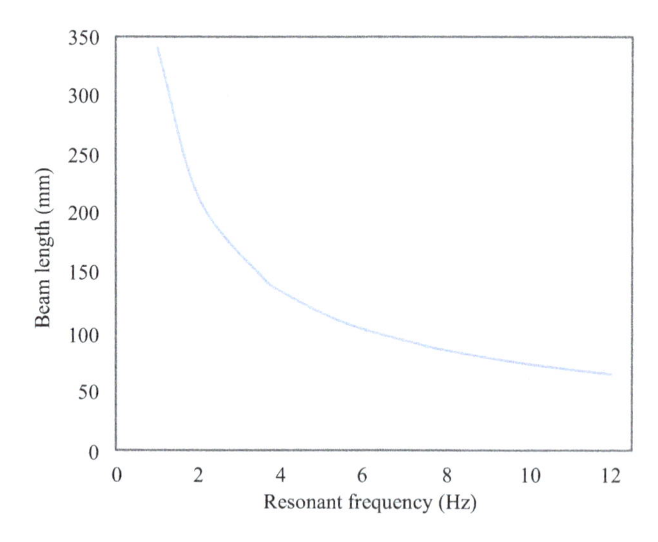


*Figure 5.2 Resonant frequency versus harvester's beam length*

and $\omega_n$ in terms of $A$ is

$$\omega^2{}_n = \left(\frac{A}{2G\xi_T}\right) \tag{5.4}$$

Considering the cantilever beam and the magnet attached as mass–spring–damper system with base excitations

$$X = \left(\frac{Y\omega^2}{\omega^2{}_n\sqrt{\left[1-\left[\frac{\omega}{\omega n}\right]^2\right]^2+\left[2\xi_T\left[\frac{\omega}{\omega n}\right]\right]^2}}\right) \tag{5.5}$$

with an amplitude of base excitations $Y$ and amplitude of velocity as

$$G = \left(\frac{A\omega}{\omega^2{}_n\sqrt{\left[1-\left[\frac{\omega}{\omega n}\right]^2\right]^2+\left[2\xi_T\left[\frac{\omega}{\omega n}\right]\right]^2}}\right) \tag{5.6}$$

The designed spring–mass structure is more sensitive to the foot strikes because of the long spiral spring and lumped central mass. On subjecting the device to an input vibrations, the spring starts oscillating vertically due to the proof mass as drive forces, resulting in an induced EMF across the upper and lower coils. The spring–mass motion system plays a key role in changing the magnetic flux over the upper and lower copper coils when the device is in operation. An exploded view of the EMIEH drawn in SolidWorks™ is shown in Figure 5.1.

The cross-sectional view of the proposed EH is shown in Figure 5.1. It is a MDOF system and each beam can be modeled as a SDOF for simplicity of equations.

## 5.3 Experimental setup

Figure 5.3 shows the general architecture of the proposed experimental setup for the characterization of the prototype. Due to body mechanics, EH will produce electrical energy by the conversion of mechanical energy [201]. A rectifier circuit will be used to convert the alternating current (AC) to direct current (DC) to be supplied to different health-monitoring sensors [202].

## 5.4 Modal analysis

The motion-induced hybrid EH is a multimodal system with various frequency modes, as shown in Table 5.2. The device was simulated for Eigenfrequency

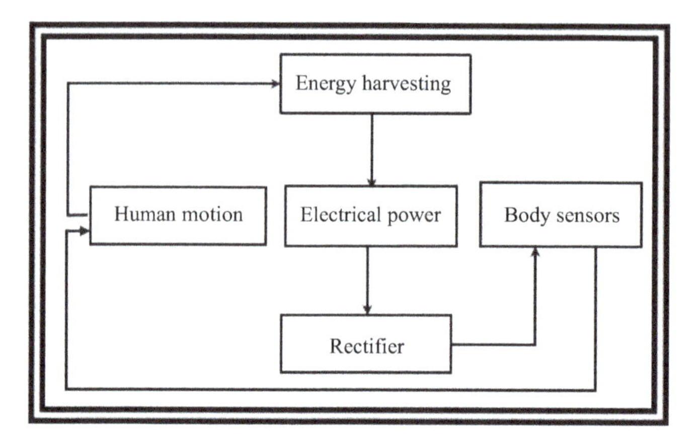


*Figure 5.3 Schematic of the proposed experimental setup*

*Table 5.2 Simulated frequencies and vibration modes of cantilever beam-coil and magnet assembly*

| **First mode** | **Second mode** | **Third mode** |
|---|---|---|
| 0.37 Hz | 0.31 Hz | 1 Hz |
| 9.8 Hz | 10.0 Hz | 11.2 Hz |
| 1.8 Hz | 2.3 Hz | 3Hz |

analysis using COMSOL Multiphysics® software. The resonant states lie within the frequency bandwidth range of the human foot strike with upper PZT beam (holding wound coil) resonance of 1 Hz, the medium beam (holding rectangular magnets) resonance as 10 Hz, and the lower movable PZT beam (on which a wound coil has been fixed on the top) showed a resonant frequency of 3 Hz.

The device consists of three different suspended systems: an upper cantilever beam holding coil, a medium beam holding two rectangular permanent magnets, and a lower beam on which a second coil has been pasted, as shown in Figure 5.1.

## 5.5 Summary

A multimodal hybrid EH is proposed which can harvest body mechanics of low-frequency oscillations. In the devised hybrid system, the upper and lower piezoelectric beams holding wound coil are actuated by the compressive force and swing motion vibrations in the tibial axis, which generates electrical power by piezoelectric conversion. A medium cantilever (GI steel) beam bearing 5 mm square magnets on the top and bottom as the tip mass can generate electrical power by electromagnetic induction according to Faraday's law of electromagnetic induction. The resonant frequencies of the hybrid energy-harvesting system are in accordance with the practical operating bandwidth (1–11 Hz) of foot-induced low-amplitude oscillations. We are working on a spiral beam introduction for the harvester to achieve resonance at very low-frequency oscillations.

# Chapter 6
# Fabrication and characterization of nonlinear multimodal electromagnetic insole energy harvesters

## 6.1 Introduction

Technological advances have made human life more intelligent and comfortable, and the past two decades have witnessed the increasing applications of healthcare-monitoring sensors, wearable electronics, portable devices, mobile terminals, and wireless transport systems that bring about more concerns on the rising needs for energy supply [203–205]. Currently, these sensors and electronics rely on either rechargeable or replaceable batteries for operation. However, conventional electrochemical batteries and microfuel cells due to the requirement for periodic recharging, limited lifetime, and environmental pollution problems are expensive, inconvenient, and tedious. Even sometimes, especially for applications in a remote and embedded environment, recharging or the replacement of batteries is made impossible due to accessibility [206]. Due to the huge potential of ambient kinetic energy sources in providing sustainable power supply [207], research efforts on vibration-to-electricity conversion technologies [75,208] have been increasing, potentially turning implantable and wearable biomedical devices into autonomous, self-powered systems. The various developed energy harvesters, capable of capturing energy from mechanical vibrations [209], bridge vibrations [180], vehicles [210], machine vibrations [211], ocean waves [212], wind [213], human motion [214], etc., are potential power sources for sensors and microelectronic devices. Currently, the most commonly used strategy for powering wearable electronics is through biomechanical energy harvesting [215], with biomechanical energy being converted into electrical energy by different mechanisms, including but not limited to EM [216], PE [217], electrostatic [218], and triboelectric [219] transductions. EM-type biomechanical energy harvesters based on rotational motion [220], rotary–translational motion [221], magnetic springs [138], rolling magnets [137], curved harvesters' design [222], magnetic levitation [223], and cycloid-type [224] approaches have been applied to wearable smart gadgets. Among the PEEHs, FUC [225], force amplification method [15], impact-driven [122], and flexible crystalline [226], etc. have previously been reported to supply energy to low-power wearable electronics. Newly introduced triboelectric nanogenerators are increasingly used for autonomous body sensors by the conversion of human motion energy into electrical energy [227]. Moreover, the integration of two or more of these technologies, such as PE–EM [108], EM–triboelectric [159], PE–triboelectric [228], and triboelectric-PE–EM [45,158] in a single hybrid device, has been recently developed to attain much higher power densities. EMEHs are extensively used and considered as a promising renewable energy source for future low-power wearable electronics, in contrast to TEHs and PEHs, which have extremely large internal impedances [199]. The higher internal impedance of PE and triboelectric materials inevitably leads to a drop in total output power and results in limited output current levels across the circuit [5]. EMEHs, on the other hand, exhibit simple structures and based on EM induction generate relatively high current and low-power levels due to the low internal impedances in these harvesters [229]. An EMEH is mainly comprised of a coil-and-magnet arrangement, with power being generated as a result of their relative movement. Depending on the designs, possible arrangements of the coil and magnet can either be a moving magnet and fixed coil, a moving coil, and a fixed magnet, or both the moving coil and moving magnet. From the comparison of the EMIEHs in Table 6.1, it is worth mentioning that the generated power of these harvesters varies from tens of $\mu$W to 0.5 mW, and the integration of the harvester with the shoe remains rather limited. Furthermore, decreasing the resonant frequency of IEHs to low-frequency ($\leq$ 10 Hz) human motion for peak output power generation and multiresonant frequencies is desirable in future design and optimization.

A centimeter-scale spiral spring-type EMIEH is reported in this chapter. The harvester introduces a circular spiral spring holding disk magnets above and below the central platform as the tip mass to respond to walking forces and attain the lowest possible resonant states. The harvester shows higher-sensitivity to low-frequency external vibrations than those of the conventional cantilever-based design, and hence allows low impact energy harvesting such as harvesting energy from walking, running, and jogging. The experimentally tested four resonant frequencies occurred at 8.9, 28, 50, and 51 Hz. At the first resonant frequency of 8.9 Hz at a base acceleration of 0.6 g, the lower EM generator can deliver a peak power of 664.36 $\mu$W and an RMS voltage of 170 mV to a matching load resistance of 43.5 $\Omega$. The upper EM generator can contribute an RMS voltage of 85 mV, corresponding to a peak power of 175 $\mu$W across 41 $\Omega$ under the same experimental conditions. The developed harvester exhibits multiresonant frequencies and can operate in a wide operating frequency range of about 41 Hz. The fabricated prototype was incorporated into the sole of a commercial shoe and produced sufficient power levels from low-frequency biomechanical vibrations and may be used as a potential power source to operate low-power electronic gadgets such as a pedometer and smartwatch. Moreover, the chapter investigates the influence of input vibrations intensity, Eigenfrequency analysis, and a model for the EMIEH is experimentally validated. Furthermore, the performance of the developed EMIEH is compared with the state-of-the-art insole and human motion-based EMEHs.

The chapter is organized as follows. Design and modeling is discussed in Section 6.2 with schematic and finite element modeling. Prototype fabrication and the experimental setup are discussed in Section 6.3, with experimental results being

*Table 6.1 Performance comparison of the reported PE and EM walking energy harvesters with EMIEH*

| Mechanism | Dimensions ($cm^3$) | Acceleration (g) | Operation freq. (Hz) | Matching load (Ω) | Power ($\mu$W) | Power density ($\mu W/cm^3$) | Power density/acceleration ($\mu W/g \cdot cm^3$) | Ref. |
|---|---|---|---|---|---|---|---|---|
| EM | 53.38 | 0.04 | 8 | 800 | 14.55 | 2.85 | 71.25 | [230] |
| EM | 7.16 | 2 | 4.6 | 17 | 110 | 15.4 | 7.7 | [231] |
| EM | 5.18 | 1.56 | 6.7 | 10 | 569 | 109.84 | 70.41 | [232] |
| EM | 19.2 | 2 | 5.8 | 85 | 103.5 | 5.4 | 2.7 | [233] |
| EM | 20.1 | — | 1 | 240 | 61.3 | 3.04 | — | [234] |
| EM | 24.5 | 0.4 | 7.4 | 185 | 263 | 10.73 | 26.82 | [235] |
| EM | 2.12 | 1 | 10 | 240 | 5.8 | 2.85 | 2.74 | [236] |
| EM | 47.8 | 0.6 | 8.9 | 41 and 43.5 | 839 | 17.56 | 29.25 | [237] |

presented in Section 6.4. The developed harvester is compared and discussed in Section 6.5. Finally, the chapter concludes with chapter summary in Section 6.6.

## 6.2 Design and modeling

### 6.2.1 Architecture and the working mechanism

Figure 6.1(a) shows the cross-section of the upper and lower EM units in the proposed moving magnet-type EMIEH. The harvester is composed of a circular spiral spring with proof mass as magnets, and wound coils located on the upper and lower part of the device. The circular spiral spring is encapsulated between upper and lower Teflon spacers, allowing it to oscillate freely. The spring is designed in a spiral geometry, allowing the maximization of its length (441 mm), in order to

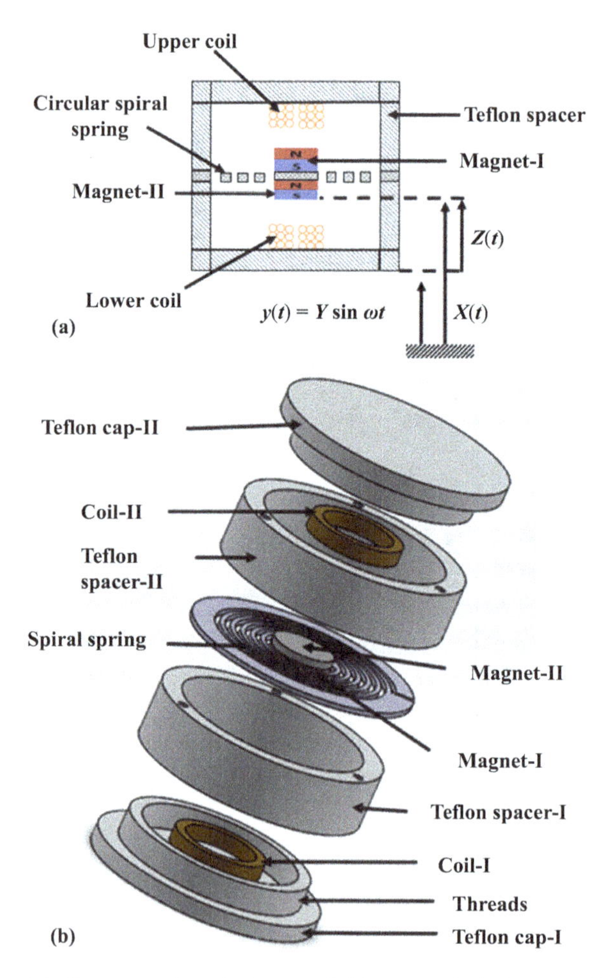


*Figure 6.1 (a) Cross-sectional view and (b) an exploded view of the EMIEH*

reduce the resonant frequency of the harvester within the range (0–11 Hz) of human step frequency [159]. Two disk magnets (Ø 12 mm × 1.5 mm) are fixed on the top and bottom sides of the central platform of the circular spring, as proof mass. The proof mass (magnets) allows oscillations upon exposure of the harvester to biomechanical vibrations, generating a voltage across the upper and lower coils. A copper wire of 80 $\mu$m diameter is used to obtain maximum number of coil's turns and hence attaining sufficient magnetic flux across the upper and lower coils. Both the wound coils are fixed to the inner side of an upper and lower Teflon cap (threaded).

The designed spring–mass structure is more sensitive to the foot strikes because of the long spiral spring and lumped central mass. On subjecting the device to input vibrations, the spring starts oscillating vertically due to the proof mass as drive forces, resulting in an induced EMF across the upper and lower coils. The spring–mass motion system plays a key role in changing the magnetic flux over the upper and lower copper coils when the device is in operation. An exploded view of the EMIEH drawn in SolidWorks™ is shown in Figure 6.1(b).

### 6.2.2 Finite element modeling

The circular spiral spring holding two NdFeB permanent disk magnets at its center was simulated as a 3D model using COMSOL Multiphysics® software with the same material and geometric properties as the fabricated prototype, as shown in Figure 6.2. The first resonant mode appears at 8.9 Hz with up-and-down movement

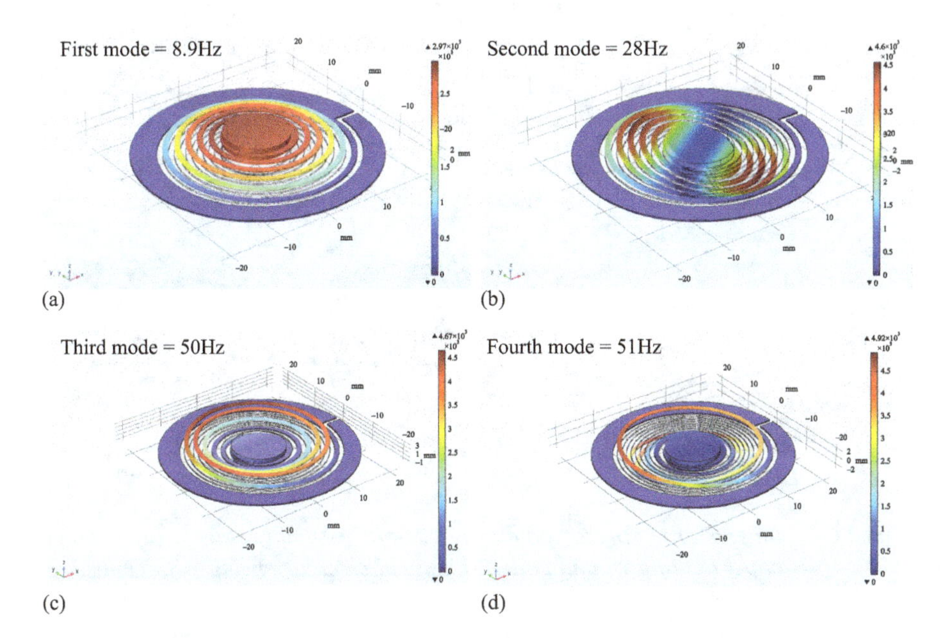


*Figure 6.2 Eigenfrequency analysis of the spiral circular spring holding disk magnets: (a) first resonance at 8.9 Hz, (b) second resonance at 28 Hz (c) third resonance at 50 Hz, and (d) fourth resonance at 51 Hz*

of the center of the spring holding magnets, as shown in Figure 6.2(a). In the second mode (at 28 Hz), the spring and the attached magnets seemed to rotate in-plane about the horizontal axis passing through the center of the magnet, Figure 6.2(b).

The third and fourth turns of the spiral spring were oscillating up and down with maximum displacing while the center seemed to be stable at the third resonant mode, as depicted in Figure 6.2(c), and the fourth resonant mode is shown in Figure 6.2(d) with the fourth turn in the spring moving. Different frequency modes of the developed EMH are shown in the video and can be accessed using the link https://bit.ly/37BoXIV.

## 6.3 Fabrication of prototypes and the experimental setup

The fabricated EMIEH is shown in Figure 6.3, which consists of an upper and lower electromagnetic generator. A lower wound coil (coil-I), having 740 turns, was bonded on the inner side of the lower threaded Teflon cap as shown in Figure 6.3(a), and an upper wound coil (coil-II) with 700 turns was securely pasted onto the upper cap. The coils were produced using a manual winding machine. The circular spiral spring was designed in PTC Creo (3D CAD). The spring was

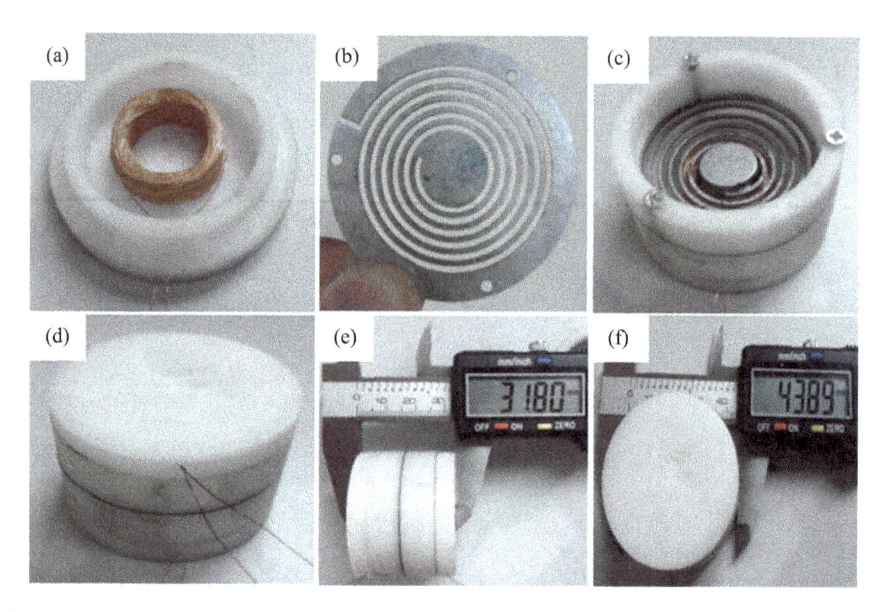


*Figure 6.3 Images of the prototype during various assembly stages: (a) Teflon cap-I holding wound coil-I, (b) spiral circular spring with central platform for the top and bottom magnets, (c) circular spiral spring carrying disk-shaped magnet-I and magnet-II on both sides is sandwiched between spacer-I and spacer-II, (d) side view of the assembled harvester, (e) height of the harvester, and (f) diameter of the EMIEH*

fabricated from commercially available galvanized steel (0.26 mm) using a computer numerical controlled, wire-cut electrical discharge machining (CNC-EDM) method, in such a way to have a maximum possible length (441 mm) in the provided constraint area of the insole. The planar spiral spring has five turns of 1 mm width with a gap of 1 mm between successive spring's turns and a circular platform of 12 mm diameter for an upper- and lower-disk magnet, as seen in Figure 6.3(b). A 10 mm high, hollow Teflon spacer having equally distant three holes of 2 mm was then worn on the lower threaded spacer, which is a base support for the spiral spring holding two 12 mm × 1.5 mm neodymium permanent disk magnets on the top and bottom sides right at the center and closed with the upper and lower wound coils. Another spacer with the same dimensions was placed over the spring (enclosing a 4 mm outer solid line) to be firmly fixed between an upper and lower Teflon spacer (Teflon spacer-I and spacer-II) using small-sized (2 mm) nuts and bolts, which allow its free oscillation, as shown in Figure 6.3(c). The gap between the magnets and coils was kept at 4 mm for achieving better flux density. The upper threaded Teflon cap was screwed over Teflon spacer-II to complete the assembled prototype, as shown in Figure 6.3(d). Both coil-I and coil-II are aligned and held apart by 3 mm from magnet-I and magnet-II. The harvester with a height of 31.87 mm, Figure 6.3(e), and a diameter of 43.8 mm, Figure 6.3(f), can easily be integrated into the sole of a commercial shoe.

The main features and dimensions of the developed EMIEH are presented in Table 6.2.

A schematic of the experimental setup is illustrated in Figure 6.4(a) and the fabricated rig is shown in Figure 6.4(b). The prototype is mounted on a Teflon block, which is tightly fixed to the vibration shaker's table. Excitation frequencies and base acceleration levels produced by the vibration shaker can be varied using a function generator (Model: GFG 8020H, GW Instek, New Taipei, Taiwan). Sinusoidal signals coming from the function generator are amplified and regulated

*Table 6.2 Prototype dimensions and properties*

| Harvester's property | Units/sizes |
|---|---|
| Magnetic flux density | 1.32 T |
| Mass of the magnet | 1.24 g |
| Diameter of the magnet | 12 mm |
| Thickness of the magnet | 1.5 mm |
| Width of the spring's turn | 1 mm |
| Thickness of the spring's turns | 0.26 mm |
| Length of the spring | 441 mm |
| Young's modulus of GI steel | 200 GPa |
| No. of turns in coil-I | 700 |
| Impedance of coil-I | 41 Ω |
| No. of turns in coil-II | 740 |
| Impedance of coil-II | 43.5 Ω |
| Gap between the coil and magnet | 4 mm |
| Overall size of the harvester | Ø 43.8 mm × 31.8 mm |

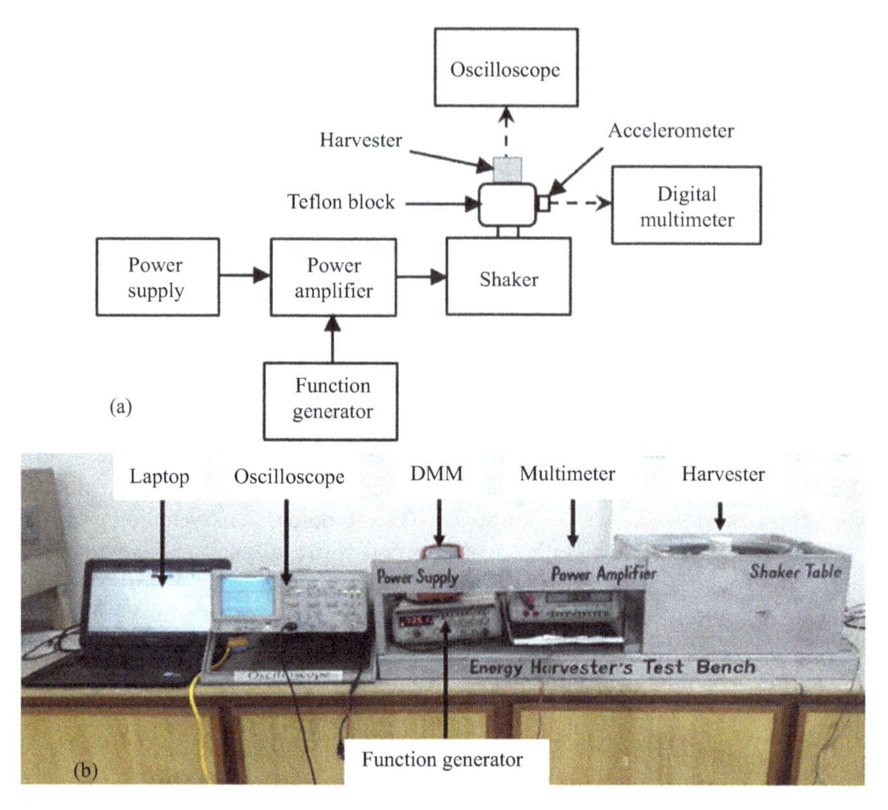


*Figure 6.4 (a) A schematic of the experimental setup and (b) a fabricated setup for the characterization of EMIEH*

using a power amplifier (Model RM-AT2900, Rock Mars, United Arab Emirates) before being supplied to the vibration shaker. An accelerometer (Model: EVAL-ADXL335Z, Norwood, MA, USA) is attached to the shaker's table, so as to measure the acceleration levels under which the harvester is subjected to. Additionally, output voltage signals across the coils and acceleration amplitudes are measured and analyzed using an oscilloscope (Model GOS 6112, GW Instek, New Taipei, Taiwan) and digital multimeters (Model: UT81A/B, Uni-Trend Technology, China), respectively. Power to the amplifier is supplied by 12 V, DC power supply (Model: GT-41132, GlobTek, Inc., Japan) during experimentation.

## 6.4 Experimental results

The performance of the EMIEH has been analyzed using a vibration shaker in the developed experimental setup. The harvester was tested for frequency response at low base acceleration levels of 0.1–0.6 g under a forward frequency sweep from 1 to 100 Hz. The prototype exhibits multiresonant frequencies with peak outputs at 8.9, 30, and 50 Hz and due to the nonlinear effects, the resonant peaks move

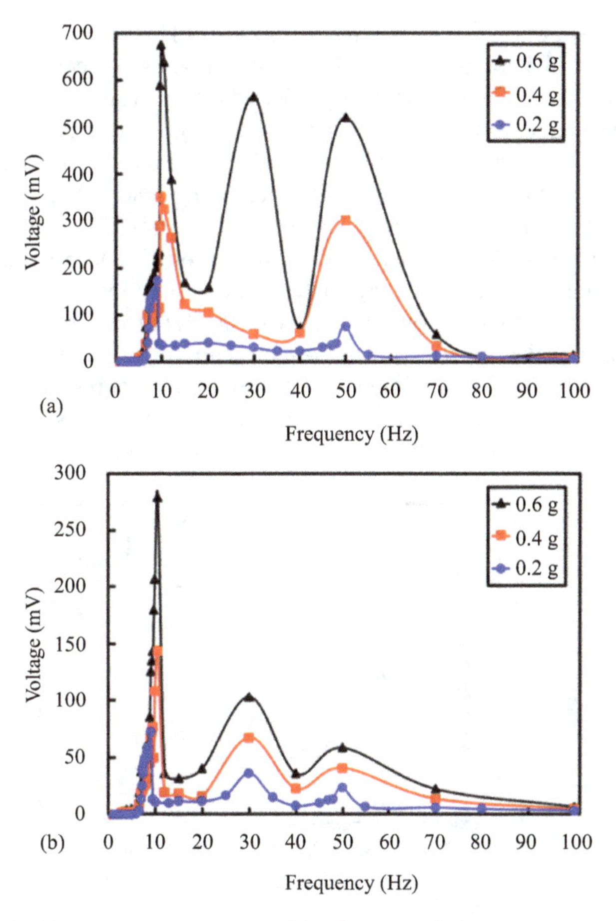


*Figure 6.5 (a) Frequency response of the lower coil for open-circuit and (b) the frequency response of the upper coil at no-load test*

slightly at relatively higher base acceleration levels as compared to the low base acceleration levels (from 7.8 Hz at 0.2 g to 8.9 Hz at 0.6 g), as shown in Figure 6.5 (a). A maximum open-circuit voltage of 675 mV was generated across the lower coil at the first resonant frequency of 8.9 Hz under 0.6 g acceleration. The peak voltage levels obtained at the second and third resonant frequencies under the same experimental conditions were 565 and 520 mV, respectively. The output of the lower EM unit is shown in Figure 6.5(a). The resonances shifted to the right at higher base acceleration levels due to the nonlinear effect and almost the same response was observed but there are two peak values across the lower coil.

The first peak is observed at 8.9 Hz under 0.6 g base acceleration with an output voltage of 280 mV, the second peak of 104 mV at 30 Hz, and the third peak at 50 Hz as 59 mV. Harmonic frequencies of the harvester are shown in the video and can be accessed using the link https://bit.ly/2Sx7A7R.

Figure 6.6 shows the frequency response (0–100 Hz) of the harvester, when an optimum load of 43.5 Ω was connected to the coil-II at base accelerations of 0.2, 0.4 and 0.6 g. As shown in Figure 6.6(a), as compared to the open-circuit voltage levels, a decrease in generated voltage levels is observed, which is due to the current drawn across the optimum load resistance; however, the maximum voltage values appear at almost the same resonant states. The highest optimum load voltage obtained across coil-I at the first resonant frequency of 8.9 Hz under 0.6 g is 121.8 mV. Moreover, the harvester generates 46.43 mV at input frequencies of 30 and 50 Hz, respectively. Figure 6.6(b) illustrates the frequency analysis at optimum load resistance connected to the upper coil under a vibration frequency (0–100 Hz) of varying amplitudes at 0.2, 0.4, and 0.6 g. For the upper coil, a maximum load voltage of 100 mV at the first resonance of 8.9 Hz under 0.6 g is generated. Furthermore, at resonant frequencies of 30 and 50 Hz, the maximum voltages are 43.2 and 16.2 mV, respectively, across the matching load resistances. Under 0.4 g, the harvester resonates at 8.7 Hz with a

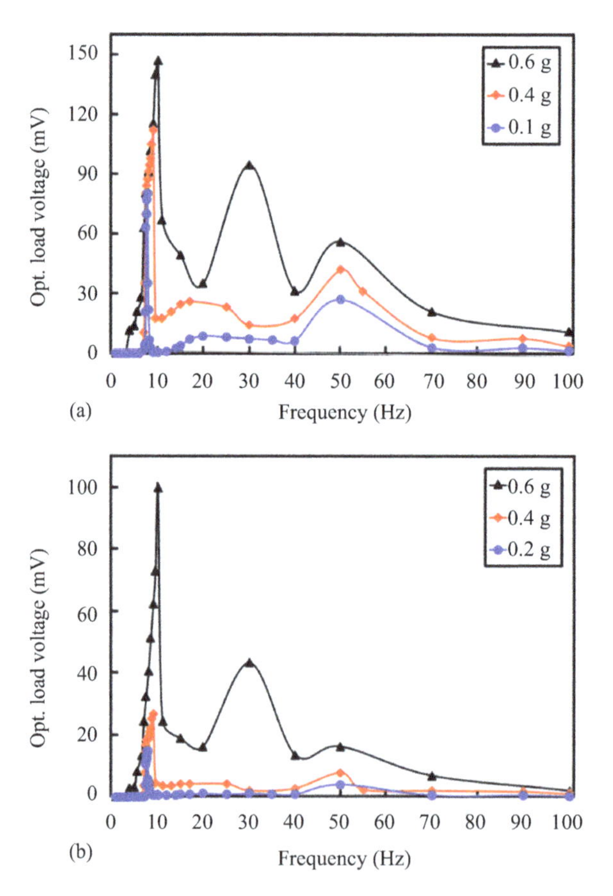


*Figure 6.6 (a) Frequency response of the lower coil across an optimum resistance and (b) the frequency response of the upper coil across an optimum resistance*

maximum load voltage of 18 mV, while under 0.2 g it resonates at 7.8 Hz with 13.7 mV load voltage. Due to the effect of gravity, magnet-I is slightly closer to the lower coil as compared to magnet-II to the upper coil, and hence more voltage is produced across the lower coil than that across the upper coil.

The RMS voltage and average power relationship under varying load resistance (1–100 Ω) across coil-I are shown in Figure 6.7(a). As the developed energy harvester is resonant type, it delivers power at resonance across the matching impedance. The RMS voltage increases with an increase in load resistance under constant vibration frequency (resonant frequency = 8.9 Hz). An RMS voltage of 195 mV was generated across 100 Ω load under 0.6 g at the first resonance. The average power of the energy harvester reached a maximum value of 664.3 $\mu$W at a resonant frequency of 8.9 Hz across an optimum load resistance of 43.5 Ω when the harvester was exposed to a base acceleration of 0.6 g. Overall, the average output power of the lower EMIEH under different load resistance ranges from 5 to 100 Ω, as shown in Figure 6.7(a). Under a low base acceleration of 0.4 g, the harvester

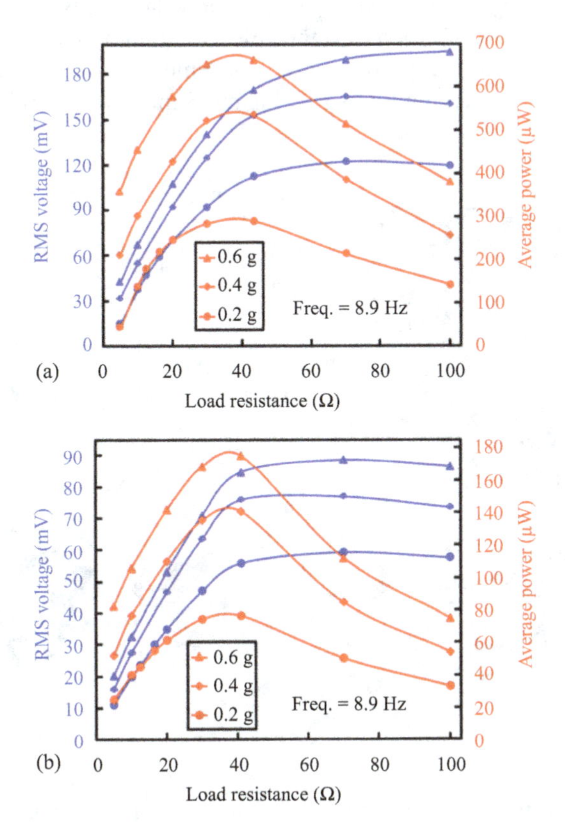


*Figure 6.7 (a) RMS voltage and average power across coil-I with respect to the load resistance at 8.9 Hz under 0.2, 0.4, and 0.6 g acceleration and (b) RMS voltage and power dependence on the external load resistance across coil-II at 8.9 Hz under 0.2, 0.4, and 0.6 g acceleration*

delivered average power of 534.6 and 291 $\mu$W under 0.2 g base acceleration across coil-I.

The dependence of the harvester output on an external load was investigated using different resistance values across the upper electromagnetic generator. Figure 6.7(b) shows the measurement results in the vibration shaker test under different base accelerations at a constant vibration frequency (8.9 Hz). The magnitude of the external resistances was varied from 5 to 100 $\Omega$ in order to study the relationship between the RMS voltage and average power, as shown in Figure 6.7 (b). Figure 6.7(b) shows that the RMS voltage increases with an increase in external load resistance, while the current drop across the generator decreases due to the ohmic losses. An average power of 175 $\mu$W was obtained across a 41 $\Omega$ load at a resonant frequency and under 0.6 g acceleration. Moreover, an RMS voltage of 86.4 mV was produced across a 100 $\Omega$ load resistance supplied under the same experimental conditions. Under low base accelerations of 0.4 and 0.2 g, the harvester produced RMS voltages of 74 and 58 mV across 100 $\Omega$ and the average power of 141 and 76 $\mu$W across the matching impedance, respectively.

The harvester was installed into the sole of a commercial shoe, as shown in Figure 6.8(a), and its output performance was demonstrated by charging a 100 $\mu$F capacitor up to 1 V with a normal walk of about 8 min, as shown in Figure 6.8(b).

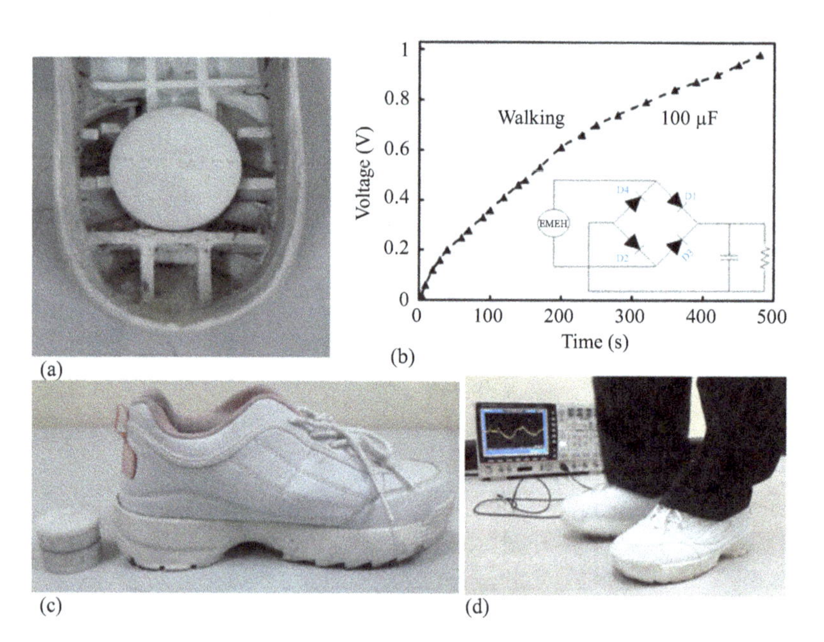


*Figure 6.8 (a) Photograph of the EMIEH integrated into a commercial shoe sole, (b) charging voltage across a full-wave bridge rectifier circuit, (c) EMIEH as a power source for wearable microelectronics, and (d) demonstration of the generated voltage using an oscilloscope when the footstep falls*

The EMIEH developed in this chapter can be a potential power source for micropower electronic gadgets, as shown in Figure 6.8(c). The device was successfully installed into the shoe and it has converted biomechanical energy into electrical energy, as shown in Figure 6.8(d). A demonstration of the generated voltage from the foot strike is shown in the video clip (https://bit.ly/2SMc7Cg).

## 6.5 Comparison and discussion

A comparison of the aforementioned biomechanical energy harvesters' parameters is summarized in Table 6.1. The resonant frequencies and base acceleration of the EMHs range from 1 to 9 Hz and 0.5–2.5 g, respectively. PIEHs, however, lie in the higher frequency range (5.6 Hz–20 kHz) due to the PE properties of these harvesters. EMEHs produced more power (61.3–10 400 $\mu$W) than that of PEHs (43–51 $\mu$W) due to the lower internal impedance (10–104.7 Ω) in comparison to piezoelectric generators with internal impedance ranging from 150 Ω to 20 kΩ.

Multiresonant states (8.9, 28, 50, and 51 Hz), compact size (47.8 $cm^3$), good power density (15.76 $\mu W/cm^3$), and power density per acceleration (26.26 $\mu W/g{\cdot}cm^3$) values of the EMIEH demonstrate that the device can provide a sustainable power solution to low-power body sensors.

## 6.6 Summary

A resonant-type EMIEH was developed and successfully demonstrated to sustainably power up low-power portable and wearable electronic devices for biomedical and human body monitoring applications. The harvester has a small dimension of Ø 31.8 × 43.8 mm and a lightweight of 47 g, which is suitable for integration into the sole of a shoe. In the device, the circular spiral spring is sensitive to the low-frequency vibrations of walking, running, and jogging due to its long length and lumped proof mass (magnets). Due to the oscillating magnets, a voltage is generated across the upper and lower coils upon subjecting to base acceleration. Moreover, the device performance has been measured for both the upper and lower coils. The device performed well over a broadband frequency range from 5 to 50 Hz. The harvester was tested using a vibration shaker inside the lab and installed into the sole of a shoe and produced sufficient power levels to operate microelectronic devices. At the first resonant frequency of 8.9 Hz and under 0.6 g, an average power of 664.3 $\mu$W and an RMS voltage of 170 mV across a 43.5 Ω load from the lower coil were obtained. The harvester produced a peak power of 839 $\mu$W from the combined output of the upper and lower coils and charged a 100 $\mu$F capacitor up to 1 V at normal walking for about 8 min.

*Chapter 7*

# Design, modeling, fabrication, and characterization of a hybrid piezo-electromagnetic insole energy harvester

## 7.1 Introduction

In the current era of IoT, there has been a shift towards wearable electronics to meet the requirement of modern living such as smartwatches, motion tracking, smart training shoes, structural, and medical health-monitoring applications [238–240]. Resource sharing and information collection are sped up, and telemedicine is now possible due to the exciting advancement in wearable and implantable sensor technologies for monitoring the critical body parameters [241]. An essential requirement for WSN operation is the power supply that can easily be harvested from human locomotion [242]. The currently used Li-ion batteries due to their finite lifespan repeated recharging, and ultimate disposal pollution problems are the technical bottlenecks towards using conventional electrochemical batteries [243]. Among the available energy sources in nature, mechanical energy from moving objects, such as machines, vehicles, bridges, and body movements in the form of walking, running, and jogging, are promising candidates for energy-harvesting applications. Technical efforts are being dedicated towards the development of microsystems in the last two decades that led the research community to fabricate self-powered sensing systems and commercial miniature devices suitable to harness the energy from body kinematics, and thus power up body sensors for monitoring vital medical signs [216,226,244,245].

Different mechanisms to harvest biomechanical energy into electricity by PE conversion [246] include low-frequency cantilever-type [247], compressive-mode [248], and frequency-tuning based [249]. Magnetic levitation-based [250], magnetic-spring-based [138], and magnetic-torque-based [251] mechanisms are the techniques among EMEHs [252]. Multiple TEEHs [253–255] have also been attempted to harness human kinetics into useful electrical energy for powering microelectronics. However, these energy harvesters still suffer from shortcomings in practical applications and need to be rectified. A considerable amount of energy is wasted usually in heat dissipation and mechanical deformation in PEEHs [253]. In EMEHs, a vibrating cantilever or a spring may end up the harvester's life due to mechanical fatigue after a long-term operation in a vibration environment [256]. Portable TEEHs need to be tuned to operate under low-frequency human vibrations [257]. Any standalone energy harvester among the aforementioned harvesting mechanisms is unable to address the above limitations and their application is restricted in network technology. Therefore, researchers are now looking into combining more than one harvesting mechanism in a hybrid harvesting system [258], being able to respond well at low-frequency body mechanics, operating at wide frequency bandwidths and have complimentary output at multiple resonant states [259].

Combining EM induction with PE transduction in a hybrid harvester has been increasingly explored by researchers due to the efficient energy conversion mechanisms [260], higher flexibility to ambient vibrations [261], longer lifecycles [230], large force capacity [262], easy implementation [217], and higher power densities [263]. Various low-frequency hybrid PE–EM energy harvesters have been proposed till date. Edwards *et al.* [264] proposed and validated experimentally a novel contact-based FUC-type HEH. Toyabur *et al.* [265] presented a multimodal energy harvester with a peak power of 250.23 $\mu$W from PE and a peak power of 244.17 $\mu$W from the EM part of the HEH. Rajarathinam and Ali [155] fabricated a HEH comprising a PE cantilever and spring–magnet arrangement. The magnet was hung on to the beam as tip mass through the spring at the free tip. Li *et al.* [266] presented a compact-sized low-frequency hybrid PEM-IEH based on a truss-and-stopper approach that produced 32 mW electrical power at a resonant frequency of 6.9 Hz under 0.7 base acceleration. Fan *et al.* [43] investigated a bidirectional hybrid PEEH for low-frequency mechanical energy environments such as biomechanics. Hamid and Yuce [267] developed a HEH that can be worn on the wrist and feet. The prototype produced an average power of 50–130 $\mu$W from different intensities low-frequency human movements by combined PE–EM transduction. Fan *et al.* [42] designed a hybrid biomechanical energy harvester that generated a peak power of 230 $\mu$W under low-frequency (5.8 Hz) input vibrations. The reported harvesting mechanisms use pollutant-free materials in converting kinetic energy from the ambient environment to sustainable power wearable microelectronics. The practice shows that the integration of HEHs to health-monitoring sensors can effectively overcome the shortcomings of traditional standalone PEHs and EMEHs. However, regardless of their proven advantages, integration of HEHs to wireless monitoring sensors is still at the onset.

Herein, a hybrid PEM-IEH is reported, which is capable of converting walking energy into electricity by a supplemental conversion mechanism, being able to power wireless sensors used in medical health monitoring for human well-being. The fabricated prototype is quite compact with a size of 46.8 $cm^3$ and a lightweight of 43.3 g. Response to low-frequency excitations, nonlinear behavior, multi-resonant states, and combined PE–EM conversion are the key factors that enhance the device performance in capturing more energy. The hybridized harvester was installed inside a commercial shoe to harvest low-frequency walking energy into electricity. The generated power can be used for use to power up small electronic devices, such as a health-monitoring sensor or a fitness tracker.

The chapter is organized as follows. Design and modeling of the hybrid PEM-IEH is elaborated in Section 7.2, with subsections discussing on the structural design, finite element modeling, and electromechanical modeling. The experimental setup and procedures are presented in Section 7.3, with experimental results being analyzed in Section 7.4. The comparison and discussion are presented in Section 7.5. Finally, the chapter concludes with Section 7.6.

## 7.2 Design and modeling

### *7.2.1 Structural design*

The schematic of the proposed hybrid PEM-IEH is presented in Figure 7.1. As shown in the cross-section, the harvester is structurally symmetrical between the

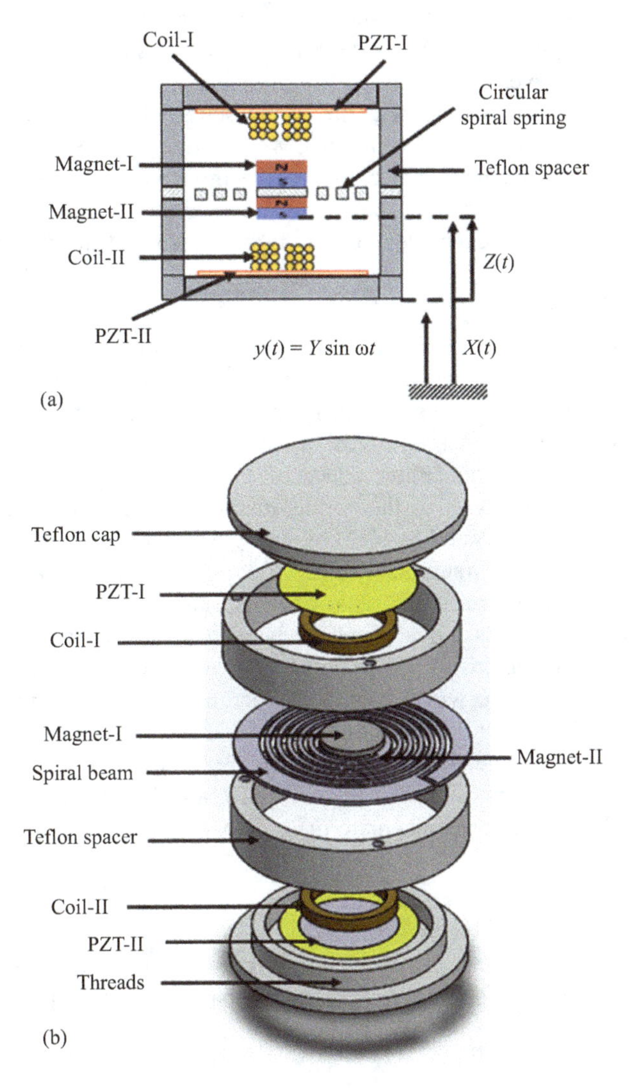


*Figure 7.1 (a) Cross-sectional view of the proposed PEM-IEH and (b) an exploded model of the PEM-IEH in SolidWorks$^{TM}$*

top and bottom. It consists of four generators: an upper PE, upper EM, a lower PE, and a lower EM generator. The upper PE (lead zirconate titanate; PZT) plate is securely bonded to the inside of an upper threaded Teflon cap, and a wound coil is pasted on this PZT plate to ensure dual transduction. A similar circular PZT is pasted onto the lower threaded Teflon holding and a wound coil is attached to the PE plate. An intermediate spiral circular spring is carrying small-sized disk magnets on its top and bottom sides on a central platform. The spring is firmly fixed between upper and lower circular Teflon spacers at three equally distant points via 2 mm nuts and bolts to move freely with input base excitations. The spring–magnet assembly is the key vibrating element due to the mass of the central magnets. The resonant frequency of the spring is decreased by increasing its length with spiral geometry for the harvester to excite at low-frequency insole vibrations (0–10 Hz). Upon exposure to input excitations, the spring starts vibrating, and the harvester generates electricity across the coils by EM induction as a result of the relative motion between the magnets and coils. As the input vibrations increase, the PE plates produce electricity by piezoelectricity and more biomechanical energy is converted into electrical energy due to the hybrid piezo-EM transduction.

### *7.2.2 Finite element modeling*

The circular spiral spring carrying disk magnets on its top and bottom sides on the central platform was designed in PTC Creo™. A 3D model with the same geometric and material properties as designed was simulated for Eigenfrequency analysis using COMSOL Multiphysics® software (Solid mechanics). Fixed constraints were applied to the outer edges of the circular spring and the model meshed. The modal analysis of the spring–magnet assembly computed four frequency modes at 8, 25, 50, and 51 Hz. The spring was excited with peak movement of the central mass at the first mode (8 Hz), as shown in Figure 7.2(a). In the second mode (at 25 Hz), the lateral inner edges of the spring moved up and down, which causes the magnetic mass moved laterally, as shown in Figure 7.2(b). The third mode (50 Hz) is characterized by the up-and-down movement of the central three turns of the spiral spring, which caused the magnets to move vertically, as illustrated in Figure 7.2(c). At the fourth mode (51 Hz), the third and fourth rings out of the five turns were moving with maximum displacement, as shown in Figure 7.2(d). Different movement patterns in the second, third, and fourth frequency modes of the spring and the associated magnetic mass displacement correspond to the peak output voltage at these modes.

### *7.2.3 Electromechanical model*

The proposed PEM-IEH can be modeled as a lumped mass linear system, with the equation of motion for a harvester experiencing base excitations [138,268,269].

$$m\ddot{z} + c\dot{z} + F_r = -m\ddot{y} \tag{7.1}$$

In (7.1), $m$ is the mass of the magnets, $z$ is the relative displacement between the magnets and coils, $F_r$ is the restoring force (spring force, $kz$) on the moving

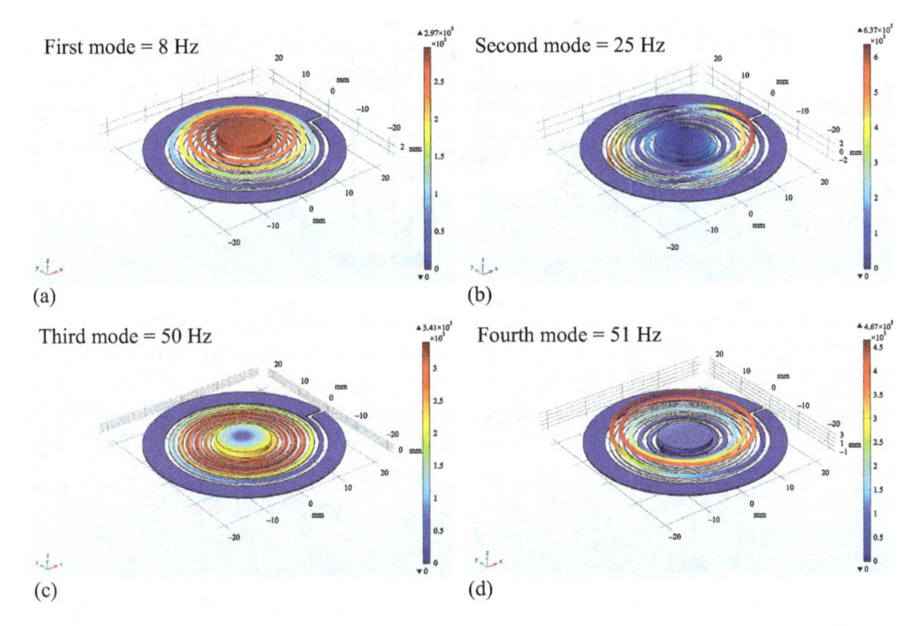


*Figure 7.2 Finite element modeling for the spring–magnet assembly: (a) first mode at 8 Hz, (b) second mode at 25 Hz, (c) third mode at 50 Hz, and (d) fourth mode at 51 Hz*

magnets, and $y$ is the displacement of the harvester's base (frame). Moreover, $c_T$ is the total damping coefficient $c_T = c_m + c_e$, which is equal to mechanical ($c_m$), and electrical ($c_e$) damping. The electrical damping coefficient $c_e$ can be expressed as $c_e = \alpha^2/R$, where $\alpha$ is the electromechanical coupling coefficient and $R$ is the sum of internal impedance and load resistance across the harvester [138].

The dynamic behavior of the spring–magnet assembly mainly depend on the magnetic mass, damping coefficient, and spring constant $k$ [206].

$$k = \frac{Gd^4}{8nD^3} \tag{7.2}$$

where $G$ is the shear stress of the spring material, $d$ is the spring sheet diameter, $n$ is the number of turns in the spring, and $D$ is the mean spring diameter.

When the harvester is exposed to base excitation $y(t)$ as shown in Figure 7.3, the relative displacement ($Z$) of magnetic masses attached to the spring may be derived in terms of the base excitation [270], as

$$Z = \frac{Y\omega^2}{\omega^2{}_n\sqrt{\left(1-\left(\frac{\omega}{\omega n}\right)^2\right)^2 + \left(2\xi_T\frac{\omega}{\omega n}\right)^2}} \tag{7.3}$$

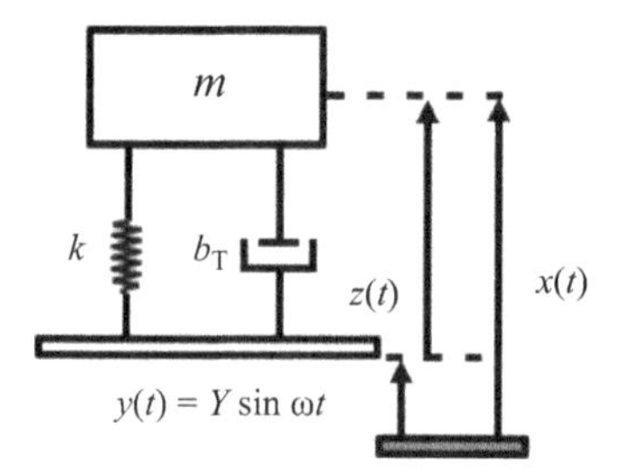


*Figure 7.3 Mass–spring–damper model of the PEM-IEH*

and the amplitude $(U)$ of the relative velocity

$$U = \frac{A\omega}{\omega^2{}_n\sqrt{\left(1-\left(\frac{\omega}{\omega n}\right)^2\right)^2+\left(2\xi_T\frac{\omega}{\omega n}\right)^2}} \tag{7.4}$$

in terms of the forcing frequency $\omega$, the input vibration amplitude $Y$, the damping ratio $\xi_T$, and the resonant frequency $\omega_n$ of the harvester [271].

The generated voltage as a function of frequency in terms of the relative velocity ($G$) of the magnet, magnetic flux density ($B_z$), and area ($S$) of coil turns can be expressed [272] according to Faraday's law of electromagnetic induction

$$V(\omega) = -G\frac{dB_z}{dx}S \tag{7.5}$$

The magnitude of EMF across the coils is directly proportional to the change of magnetic flux through the coil [252]

$$\mathcal{E} = \frac{Y}{2\xi\omega_n} = \frac{mY}{d} \tag{7.6}$$

where $d$ is the damping constant in the damping ratio equation. Since $\mathcal{E}$ is directly related to $m$ in (7.6), an increase in the proof mass ($m$) will result in increasing the EMF.

The magnetic flux density $B_x$, along the normal line to the center of the magnet [273],

$$B_x = \frac{B_r}{2}\left(\frac{x+H_m}{\sqrt{(x+H_m)^2+(r_m)^2}} - \frac{x}{\sqrt{(x^2+r_m{}^2)}}\right) \tag{7.7}$$

depends on the height of the magnet $H_m$, the flux density $B_r$, the radius of the magnet $r_m$, and distance $x$ from the magnet which can be modified to (7.8)

$$\frac{dB_x}{dx} = \frac{B_r}{2}\left(\left(\frac{D_1 + H_m}{\sqrt{(D_1 + H_m)^2 + (r_m)^2}} - \frac{(D_1 + H_m)^2}{\sqrt[3]{(D_1 + H_m)^2 + (r_m)^2}}\right)\right.$$
$$\left.-\left(\frac{1}{\sqrt{{D_1}^2 + {r_m}^2}} - \frac{{D_1}^2}{\sqrt[3]{{D_1}^2 + {r_m}^2}}\right)\right) \tag{7.8}$$

for multilayered wound coil with $N$ number of turns, inner radius $r_p$, and diameter $d_w$ of the wire. The area sum, $S$ can be obtained by taking derivative for $x$ and substituting its value in $D_1$ (a gap between the coil and the magnet).

$$S = \sum_{i=1}^{N} S_i \approx \sum_{i=1}^{N} \pi {r_i}^2 \tag{7.9}$$

$$r_i = r_p + \left(i - \frac{1}{2}\right)d_w \tag{7.10}$$

For a multilayered wound coil as shown in Figure 7.4, the time response of the voltage gain can be converted to the frequency domain.

$$V(\omega) = -G\sum_{i=0}^{n} \frac{B_r}{2}\left(\left(\frac{D_i + H_m}{\sqrt{(D_i + H_m)^2 + (r_m)^2}} - \frac{(D_i + H_m)^2}{\sqrt[3]{(D_i + H_m)^2 + (r_m)^2}}\right)\right.$$
$$\left.-\left(\frac{1}{\sqrt{D_i^2 + {r_m}^2}} - \frac{{D_0}^2}{\sqrt[3]{D_i^2 + {r_m}^2}}\right)\right)S \tag{7.11}$$

as a function of $D_i$, the distance of a single layer from the magnet.

For the resistor-connected circuit, the voltage across the load resistance can also be derived as

$$V_R(t) = I_R(t)R \tag{7.12}$$

peak voltage and power $P_R$ across an optimal resistance [274]

$$V_{L\text{peak}} = \left(\frac{R_L}{R_L + R_C}\right)V_{\text{peak}} \tag{7.13}$$

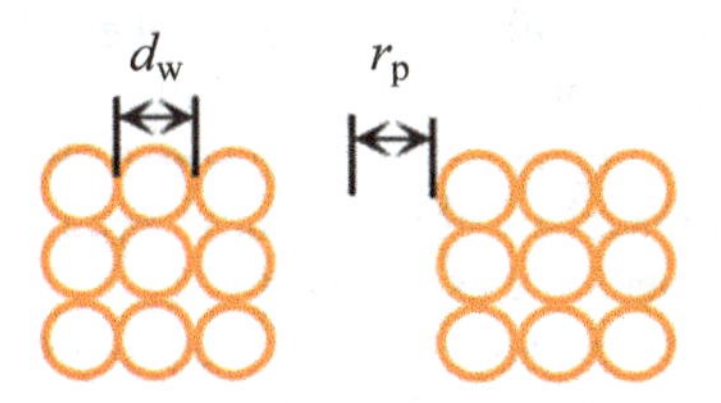


*Figure 7.4 Multilayered wound coil cross-section*

$$P_R(t) = V_R(t)I_R(t) \tag{7.14}$$

load power $P_L$ across the load resistance $R_L$

$$P_L = \frac{V_L{}^2}{2R_L} \tag{7.15}$$

depends on the load resistance $R_L$ and the coil impedance $R_C$ [275].

The maximum power output from the EM part at resonance [154] is

$$P = \frac{(NlBY)^2}{16\xi_t^2 R_c} \tag{7.16}$$

where $N$ represents the number of turns in the coil, $l$ is the length of the wire, $B$ is EM induction, $Y$ is the displacement of the spring, and $R_c$ is the coil resistance.

For a load resistance equal to the source internal resistance, connected across the PE plate,

$$V_p(t) = V_R(t) \tag{7.17}$$

$$I_R(t) = \omega Q_p(t) \tag{7.18}$$

$Q_p$ is the charge on the piezoceramic plate, and the maximum power [276]

$$P = \frac{V^2{}_L}{R_L} \tag{7.19}$$

## 7.3 Fabrication and the experimental setup

A macroscale prototype is fabricated to validate the proposed hybrid PEM-IEH model. Figure 7.5 shows the different stages during the assembly of the harvester. In the device, the outer shells including the identically-sized top and bottom caps (Ø 43 mm × 3 mm) and two circular spacers (Ø 43 mm × 10 mm) were fabricated from commercially available PTFE (poly-tetra-fluoro-ethylene) Teflon round bars (Jayant Impex Pvt. Ltd, India). The upper and lower caps hold circular PZT plates (Ø 26 mm × 0.3 mm) on its inner sides, and wound coils (Ø 19 × 4 mm), fabricated by a manual winding machine from copper wire (100 $\mu$m) are fixed to the plates, as shown in Figure 7.5(a). A circular spring (Ø 43 mm × 0.26 mm), fabricated from hot-dipped galvanized steel using CND-based wire-cut electrical discharge machining (CNC-EDM) with a length of 440 mm and 5 number of turns, is carrying NdFeB permanent neodymium disk magnets (Ø 12 × 1.5 mm) on its top and bottom sides as central masses, as shown in Figure 7.5(b). The spiral spring is screwed between the Teflon spacers using 2 mm nuts and bolts, to allow free movement when subjected to low-frequency external excitations, as shown in Figure 7.5(c). The spacers are then worn over the circular Teflon caps to complete the prototype assembly, Figure 7.5(d). In the devised harvester, the spring is the key oscillating component (mass–spring system) for generating a voltage across the coils by EM induction and PE transduction

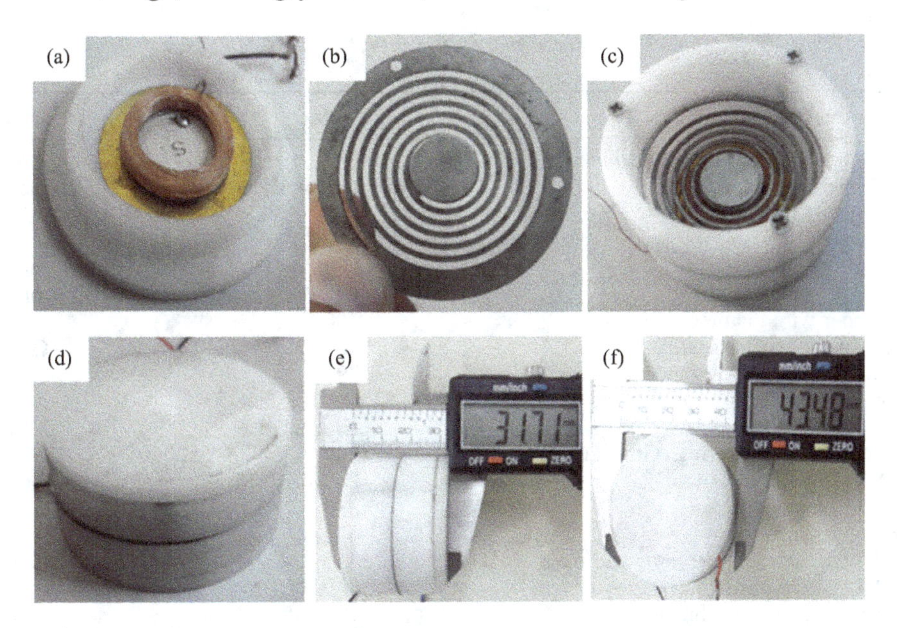


*Figure 7.5 Assembly of the fabricated hybrid prototype: (a) lower threaded Teflon cap holding circular PZT and a wound coil above it, (b) the spiral spring holding disk magnets on the top and bottom sides, (c) the spring screwed between the circular spacers, (d) side view of the assembled hybrid PEM-IEH, (e) height, and (f) diameter of the PEM-IEH*

on exposure to external vibrations. It starts oscillating at a low frequency of 3 Hz, and at higher base excitations and resonant frequencies, the magnets touch the upper and lower PZT plates and contribute in generating electrical energy across the PZT by piezoelectricity. Figure 7.5(e) depicts the height of the assembled prototype, while the diameter of the harvester is shown in Figure 7.5(f). The hybrid PEM-IEH can be easily installed into the heel of a commercial shoe.

The geometric and material properties of the fabricated PEM-IEH are listed in Table 7.1.

The schematic diagram of the experimental setup for testing the fabricated prototype inside the laboratory is shown in Figure 7.6(a). Figure 7.6(b) shows the developed experimental rig. The testing rig consists of a vibration shaker, function generator, amplifier, oscilloscope, multimeters, a 3-axis accelerometer, DC power supply, and a computer. The harvester was fixed on the shaker's table (made of Teflon bar and fixed to the shaker). The vibration shaker is capable of producing excitations with different frequencies and base accelerations. The intensity of base accelerations was recorded using an accelerometer (Model: EVALADXL335Z, Norwood, MA, USA) fixed to the shaker's table. The function generator (Model: GFG 8020H, GW Instek, New Taipei, Taiwan) produced sinusoidal frequency signals that are magnified and regulated by the power amplifier (Model RM-AT2900, Rock Mars, United Arab Emirates) that excites the vibration shaker. The amplifier was

*Table 7.1 Parameters and geometry of the fabricated PEM-IEH*

| Parameter | Value | Unit |
|---|---|---|
| Magnet-I dimension | Ø 12 × 1.5 | mm |
| Magnet-II dimension | Ø 8 × 1.5 | mm |
| Magnet-I mass | 1.24 | g |
| Magnet-II mass | 1.12 | g |
| Coil-I dimension | Ø 19 × 4 | mm |
| Coil-II dimension | Ø 15 × 4 | mm |
| Copper wire diameter | 100 | $\mu$m |
| Coil-I internal resistance | 48.5 | Ω |
| Coil-II internal resistance | 39 | Ω |
| Spring dimension | Ø 43 × 0.26 | mm |
| Spring material | GI steel | N/A |
| Modulus of the spring (GI steel) | 200 | GPa |
| PZT dimension | Ø 27 × 0.35 | mm |
| PZT internal resistance | 330 | Ω |
| Spring length | 440 | mm |
| The gap between the coils and magnets | 4 | mm |
| Mass of the prototype | 43.3 | g |
| Prototype overall dimension | Ø 43.48 × 31.71 | mm |

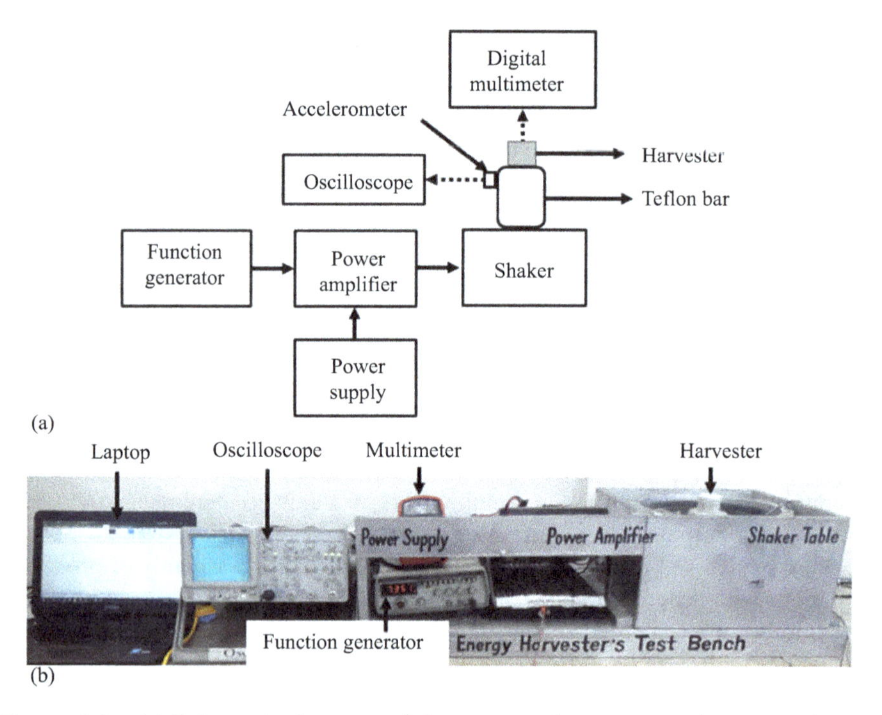


*Figure 7.6 (a) Schematic diagram of the proposed experimental setup and (b) developed testing rig*

powered using a 12 V, DC power supply (Model: GT-41132, GlobTek, Inc., Japan). Additionally, base acceleration levels from accelerometer were measured and analyzed using an oscilloscope (Model GOS 6112, GW Instek, New Taipei, Taiwan). Moreover, digital multimeters (Model: UT81A/B, Uni-Trend Technology, China) were used to measure and record the voltage output from the upper and lower hybrid PE-EM generator. The PEM-IEH was characterized for the forward frequency sweep (FFS) from 1 to 80 Hz. Since the EM and PE parts of the PEM-IEH are in-phase and connected, the combined frequency response of the PEM-IEH was tested.

## 7.4 Experimental results

The hybrid PEM-IEH was characterized for frequency response under various base accelerations (0.1, 0.3, and 0.5 g) inside a laboratory. Figure 7.7(a) shows the

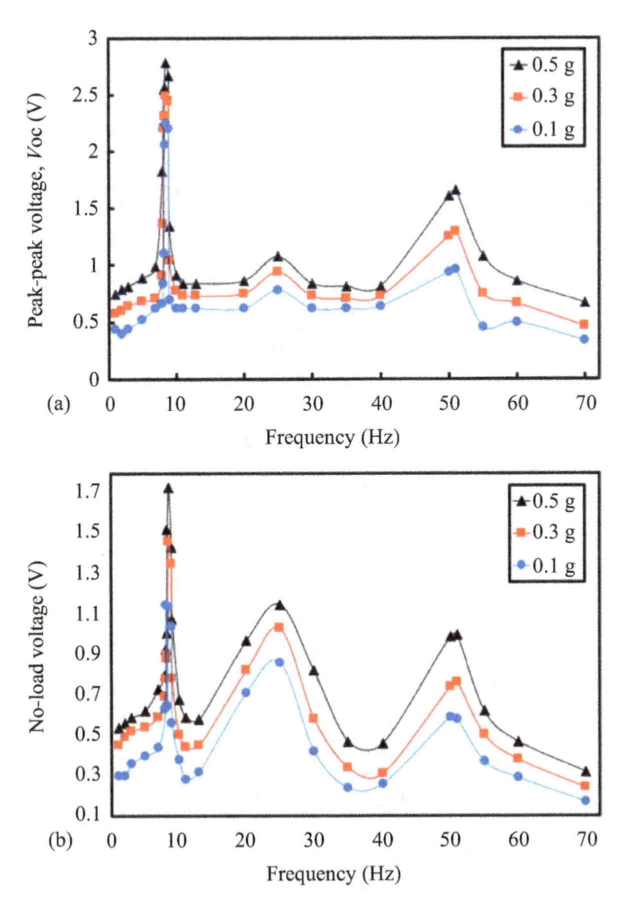


*Figure 7.7 (a) Frequency response of the upper hybrid PEM-IEH and (b) lower hybrid PEM-IEH frequency response under 0.1, 0.3, and 0.5 g base accelerations*

measured open-circuit peak-to-peak open-circuit voltage of the upper hybrid PE–EM generators, while Figure 7.7(b) shows the peak-to-peak open-circuit voltage of the lower hybrid generator. Experimental results from both figures show peak voltage values at four resonant frequencies. The resonant states occurred experimentally at 8, 25, 50, and 51 Hz. The highest open-circuit voltages 2.7 and 1.75 V were recorded at the first resonant frequency across the upper and lower hybrid units, respectively, under 0.5 g input acceleration.

Figure 7.8(a) and (b) illustrates the RMS voltage and average power as a function of load resistance across the upper and lower EM portions of the PEM-IEH, respectively. Under various base accelerations of 0.1, 0.3, and 0.5 g, different load resistances from 5 to 100 Ω were connected across the upper and lower coils to

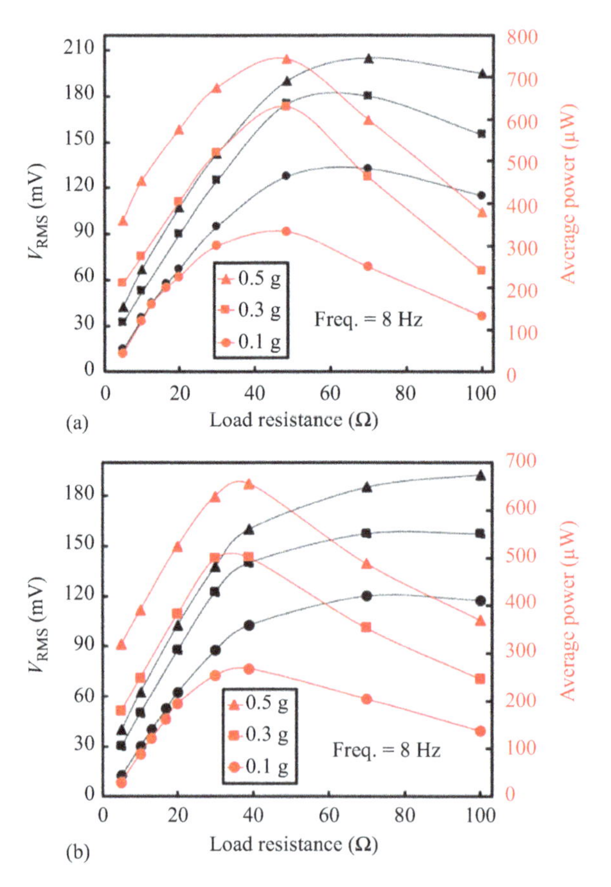


*Figure 7.8 (a) Output voltage and average power versus load resistances across the upper electromagnetic unit at 8 Hz under different base accelerations and (b) output voltage and average power versus load resistances across the lower electromagnetic unit at 8 Hz under various base accelerations*

analyze the harvester's performance at resonance. An RMS voltage of 190 mV was generated across coil-I, corresponding to an average power of 744 $\mu$W across an optimum load resistance of 48.5 Ω under 0.5 g base acceleration at the first resonant frequency of 8 Hz. The lower EM part produced an RMS voltage of 160 mV, corresponding to a 656 $\mu$W peak power across a 39 Ω optimum load under 0.5 g base acceleration at the first resonance.

The output voltage and power across the upper PZT plate are depicted in Figure 7.9(a). Different load resistances from 10 to 600 Ω were connected across the piezoceramic wafer plate under a constant base acceleration of 0.5 g at resonant

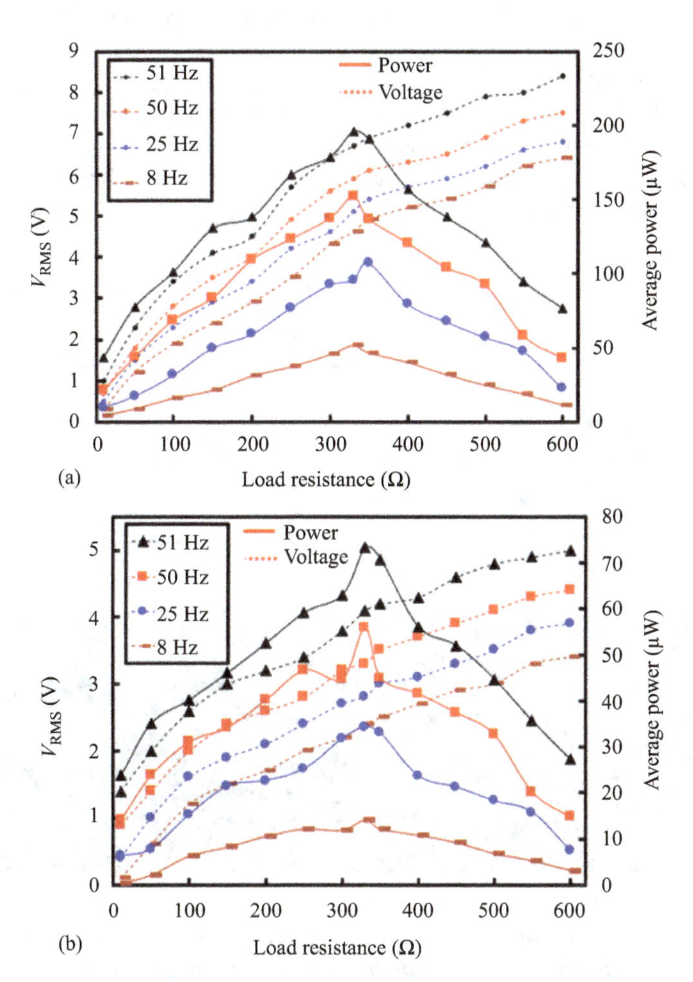


*Figure 7.9 (a) Load voltage versus load resistance from upper PZT at different resonant states under 0.5 g base acceleration and (b) load voltage as a function of load resistance across lower PZT at resonant frequencies under 0.5 g*

frequencies of 8, 25, 50, and 51 Hz. As shown in the graph (Figure 7.9(a)), the generated voltage increases with increasing input frequency, base acceleration, and resistance value across the PZT plate.

An average power of 196 $\mu$W at the fourth resonance under 0.5 g base acceleration across an optimum resistance of 330 $\Omega$ satisfies the maximum power transfer theorem. Similarly, an increase in the generated voltage is observed with an increase in base acceleration, input frequency, and an increase in the load value across the lower piezoceramic plate, as presented in Figure 7.9(b). The maximum load voltage across the lower PZT was 73 $\mu$W under 0.5 g at resonance.

The PEM-IEH was integrated into the sole of a commercial shoe, Figure 7.10 (c), for harvesting biomechanical energy from walking, running, and jogging on a treadmill, as shown in Figure 7.10. The generated voltage from both PE and EM parts fluctuated periodically due to the changing, walking, and jogging strides on the treadmill. The energy was stored in a 100 $\mu$F capacitor by connecting it to a full-bridge diode rectifier and then across the harvester, as shown in Figure 7.10(a).

By connecting the terminals of the PEM-IEH across a full-bridge diode rectifier, a 100 $\mu$F capacitor was charged up to 2.4 V, from walking for 10 min, as

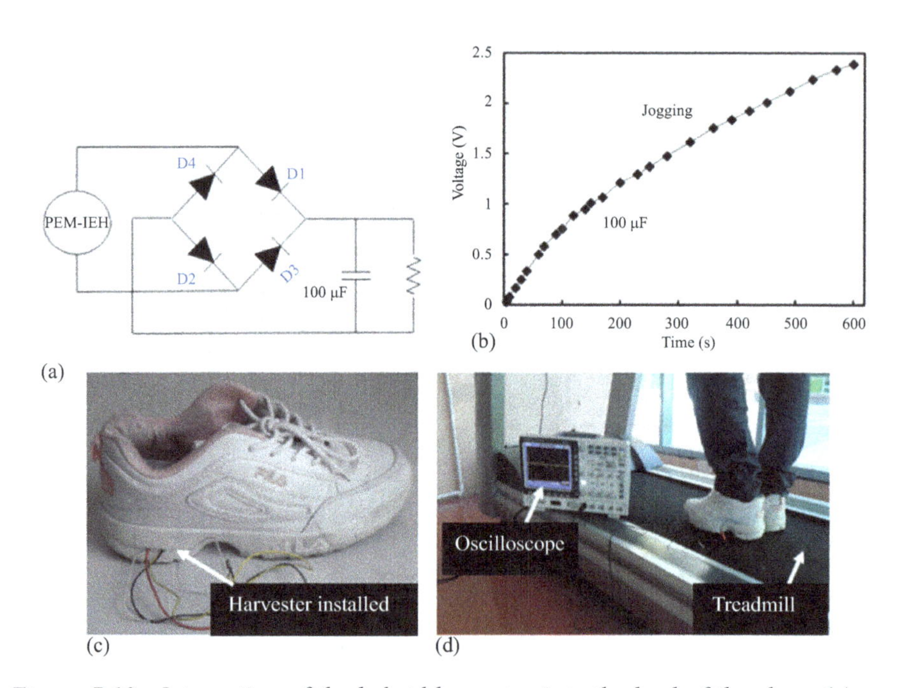


*Figure 7.10 Integration of the hybrid harvester into the heel of the shoe: (a) a full-bridge diode rectifier connected to the harvester for DC output storage into the capacitor, (b) capacitor charging during jogging on a treadmill, (c) a self-powered packaged shoe for biomechanical energy harvesting, and (d) demonstration of the electrical output from walking on a treadmill*

*Table 7.2 Comparison of the reported and developed insole energy harvesters*

| Harvesting mechanism | Device size ($cm^3$) | Optimum load (Ω) | Operation frequency (Hz) | Vibration intensity (g) | Peak power (μW) | Power density (μW/$cm^3$) | Ref. |
|---|---|---|---|---|---|---|---|
| EM | 47.1 | 12 | 9 | 0.8 | 1 150 | 24.41 | [2] |
| PE | 51.84 | 30k | 20k | 0.4 | 51 | 0.59 | [122] |
| EM | 78.5 | 5 | 9.1 | 0.85 | 420 | 5.35 | [138] |
| PE | 5 | 150k | 5.6 | 2 | 43 | 8.6 | [225] |
| PE | 576 | 3.3 M | 3 | — | 5 | — | [277] |
| EM | 6.47 | 25 and 50 | 5.1 | 2.06 | 2 150 | 78.47 | [206] |
| EM | 7.16 | 17 | 4.6 | 2 | 110 | 15.4 | [231] |
| EM | 5.18 | 10 | 6.7 | 1.56 | 569 | 109.84 | [232] |
| EM | 19.2 | 85 | 5.8 | 2 | 103.55 | 5.4 | [233] |
| EM | 20.1 | 240 | 1 | — | 61.3 | 3.04 | [234] |
| EM | 2.12 | 240 | 10 | 1 | 5.8 | 2.85 | [236] |
| Hybrid PE–EM | 46.8 | 39 and 48.5 | 8 | 0.5 | 1 670 | 35.68 | [182] |

$g = 9.8$ m/s$^2$.

shown in Figure 7.10(b). The output terminals of the prototype were connected to a digital oscilloscope for recording the voltage waveform under the walking test on the treadmill, as shown in Figure 7.10(d).

## 7.5 Comparison and discussion

A performance comparison of the work reported in this research with the previously reported macroscale PE and EMIEHs is presented in Table 7.2. The insole energy harvesters are commonly compared based on the generated power and voltage, the internal resistance of the harvester, device size, base acceleration, and resonant frequencies. In comparison with the reported harvesters, the PEM-IEH exhibits good performance in terms of the generated power, power density, and operating frequency under comparatively lower base acceleration levels. Moreover, the reported hybrid device has simple structured, compact, lightweight, with low fabrication cost and can easily be incorporated into the heel of a normal shoe. The prototype was tested for a long time inside the sole of a shoe and operated successfully with sufficient power levels at multiresonant states.

## 7.6 Summary

This chapter reports a hybrid PEM-IEH which scavenges biomechanical energy from walking, running, and jogging into useful electrical energy by PE and EM dual transduction mechanism. The harvester starts responding to very low-frequency walking steps (at about 3 Hz) due to the vibrations of the central spiral spring carrying magnetic masses and excites at 8 Hz (first resonant frequency), producing voltage by EM as well as PE transduction. Similarly, the input mechanical excitations cause strain in the PE ceramic plates, and a voltage is generated by piezoelectricity across the PE elements. The harvester excited at four different resonant states with increased bandwidth and peak power at these frequencies. The fabricated prototype was tested inside the laboratory on a vibration shaker and incorporated inside the sole of a commercial show to demonstrate its performance during walking and jogging on the treadmill. The maximum average power obtained across the upper and lower wound coils at the first resonant frequency under 0.5 g base acceleration was 744 $\mu$W across 48.5 Ω and 656 $\mu$W across 39 Ω load resistances, respectively. Peak powers of 196 and 73 $\mu$W generated across the optimum load resistance of 330 Ω across the upper and lower PZT plates at 51 Hz resonant frequency under 0.5 g base acceleration. Successful integration of the harvester into the sole and capacitor charging performance reveals its potential application in microelectronic gadgets, such as a smartwatch, a pedometer, and a wireless tracker.

*Chapter 8*

# Multi-degree-of-freedom hybrid piezoelectromagnetic insole energy harvesters

## 8.1 Introduction

Portable electronic devices, including wearable sensors, are in high demand so as to allow the continuous monitoring of physical and well-being parameters, such as heartbeat, blood pressure, diabetes, number of walking steps, and athletic activities in real time [278–280]. Conventional wearable electronics used in various fields, including but not limited to healthcare, military, academic, agriculture, consumer, environment, finance, and retail, are mainly powered with batteries, having limited operational life cycles and some associated hazards [281,282]. Battery recharge or replacement in embedded and remotely located WSNs is inconvenient and cannot satisfy the "plug and play" concept [283]. Numerous researches have focused on increasing the power density of batteries, in order to extend the battery life, as well research works are underway to reduce the power consumption of wearable sensors and portable electronic gadgets. However, usually, microsystems are still highly battery-dependent and since there is as yet no onboard self-powered mechanism, these microsystems require frequent replacement of batteries [284–286]. To overcome this limitation, a self-powered energy-harvesting system integrated into the monitoring sensor for its sustainable operation is required [287,288]. There are some possible energy-harvesting sources, such as solar, thermoelectric, and wind; however, due to the intermittent sunlight, lower body heat, and nonpersistent wind flow, the previously developed energy harvesters still need optimization to guarantee a nonstop power supply to wearable microdevices [289].

Biomechanical energy harvesting may be a potential alternative power source for smart clothing, biomedical devices, and sports apparel, as listed in Table 8.1, which may sync wirelessly to a smartphone or a watch for further transmission and processing [290]. IEHs can sustainably operate health-monitoring telemetry circuits, GPS tracking chips for hiking, low-power Bluetooth transmitters, RF, and even an Arduino microcontroller [291].

Recently, biomechanical energy harvesting has become an increasingly research-attractive interest for achieving autonomy in health-monitoring applications due to the more efficient and conveniently available energy from body kinematics and kinetics. Table 8.1 lists different wearable sports apparel. Limb movements and, mainly heel strike, which provides mechanical vibrations of

*Table 8.1 Applications of different wearable low-power body sensors*

| Device type | Physical placement | Battery required (V) | Monitoring applications | Ref. |
|---|---|---|---|---|
| Vital sign sensor | Wrist-worn | 1.8–3.3 | Heart rhythm, blood pressure, oxygen, and body temperature monitoring | [292] |
| Cardiovascular sensor | Arm or thigh | 1.5–2 | Stress evaluation by heart rate variability | [293] |
| Pedometer | Ankle | 1.5 | Step counter: measures walking speed, distance, and calories burned | [294] |
| Accelerometer | Ankle strap/ wrist worn | 1.5 | A 3-axis wireless motion tracker, seizure activity | [295] |

*Table 8.2 Literature summary of the frequency and acceleration levels at the shoe sole*

| Harvester type | Acceleration level (g) | Frequency (Hz) | Ref. |
|---|---|---|---|
| PE | 0.2–0.3 | 2–3 | [296] |
| PE | 2 | 2 | [225] |
| PE | — | 1.5 | [214] |
| PE | 1 | 1 | [297] |
| PE | 5 | — | [298] |
| PE | 4 | 1.22 | [299] |
| PE | <1 | 0.5–5 | [300] |
| EM | 0.85 | 9.1 | [138] |
| EM | 2 | 10 | [139] |
| EM | 2.06 | 5.1 | [206] |
| EM | 1 | 2.75 | [230] |
| Hybrid | — | <10 | [159] |

considerable acceleration levels and frequency content, as summarized in Table 8.2, can be harvested constantly and ubiquitously as a sustainable power supply to make wearable electronics self-powered, which is a challenging task to be solved.

Biomechanical energy conversion can be converted into electrical energy using PE [301], electrostatic [218], EM [216], triboelectric [302], or hybrid [303] transduction mechanisms. VEEHs usually generate peak power near the resonant states, thus hindering operation at wide frequency bandwidths [304]. Several methods have been reported in the design of IEHs to reduce the resonant frequency and increase the device's frequency bandwidth with the introduction of PE materials; however, in fact, due to the miniature design of insole generators, these devices resonate at much higher frequencies >100 Hz [300,305]. Flexible PE and triboelectric harvesters have been recently reported for harvesting low-frequency body mechanics and textile-based wearable nanogenerators [306], where moving charges can be induced by polarization and rubbing between an electrode and a dielectric, respectively, to bridge the frequency disparity [307]. TENGs have also been used as a sustainable energy source in active sensing and self-charging modules because of their excellent efficiency, miniature size, and lightweight [308]. However, the natural frequencies of most of the previously reported energy harvesters are still on the higher side and inevitably, perform suboptimally under low-frequency human body vibrations [258]. Zhu *et al.* [253] integrated a power-harvesting shoe insole using flexible TENGs that generated a no-load voltage of 220 V and a short-circuit current of 600 μA. The harvested energy has been used to power commercial LEDs installed in the shoe. A hybrid triboelectric-EM walking-based energy harvester embedded into the sole has been shown to produce enough power to operate a GPS or recharge a mobile phone [309]. The miniature nano-generator delivers a voltage of 13.2 V and a current of 3.02 mA and is capable of recharging a Li-ion battery from 2.62 to 3.06 V after normal walking for 30 min. A nonresonant cycloid-curved-inspired wearable EM-based energy harvester that can be worn on wrist and foot to harvest low-frequency biomechanical energy is reported in [224]. The miniature device occupies a volume of 11.97 $cm^3$ and delivers an average power of 8.8 mW to a matching impedance of 104.7 Ω in response to hand-shaking vibration frequency of 5 Hz and 2.5 g base acceleration. Moreover, an energy conversion efficiency of 7.7% and a power density of 0.73 $mW/cm^3$ were reported for the harvester. Recently, Rodrigues *et al.* [158] developed an optimized hybrid nanoharvester with multipatterned (parallel, arched, and zigzag) triboelectric plates for insole applications. The harvester produces an output voltage of 14 V at an applied foot force of 390 N.

For the fabrication of an efficient IEH to harness power from walking, running, and jumping, four important challenges need to be addressed, namely a method of reducing the resonant frequency ($\leq 11$ Hz) of the harvester to convert the intermittent energy, compact size (<5 cm × 5 cm × 3 cm), lightweight (<100 g), and successful integration of the device [159,227,310]. Lowering down the resonant frequencies, broadening the operation frequency range, compact size, lightweight, multiresonant states, multi-degrees-of-freedom, and hybrid transduction mechanism in a single system that results in enhancing the overall efficiency and optimization are still some of the challenges in footwear applications.

In this chapter, a compact, lightweight, multi-degrees-of-freedom, low-frequency resonant-type multimodal HIEH has been proposed and developed to overcome the constraints in previously reported IEHs, by introducing the square spiral planar spring for insole applications. The harvester is presented with four generators, namely an upper PE, upper EM, lower PE, and lower EM generators. The HIEH that has overall dimensions of 3.9 cm × 3.9 cm × 2.9 cm and a weight of 33.2 g has been incorporated into a commercial shoe and has been shown to maintain a stable voltage supply from 5 to 55 Hz. On connecting the hybrid harvester to a full-wave rectifier circuit, a 100 μF capacitor was charged from 0 to ~1.8 V from the PE portion during normal hand movement for about 8 min. Furthermore, the same capacitor was charged up to 2.9 V by the hybrid PE–EM coupling at the same time and has shown better charging performance than that of the individual EM or PE unit. The harvester responds well under low frequency, shows better performance in capacitor charging, and can be used as a potential source for powering small electronic gadgets.

The chapter is organized as follows. A schematic design, working mechanism, and finite element modeling of the developed hybrid device are detailed in Section 8.2. Subsequently, the prototype fabrication and the experimental setup are described in Section 8.3. Section 8.4 presents the results and discussions on experiments performed on the developed harvester. Finally, the chapter concludes with comparison and discussion in Section 8.5.

## 8.2 Design and modeling

Figure 8.1(a) illustrates the vertical cross-section of the developed HIEH, comprising a set of wound coils, a set of magnets, a couple of PVDF cantilever beams, and a square spiral planar spring (steel) in the middle of the device. An exploded view of the HIEH is depicted in Figure 8.1(b). In the device, the middle portion (platform) of the square spiral planar spring holds onto two permanent magnets as

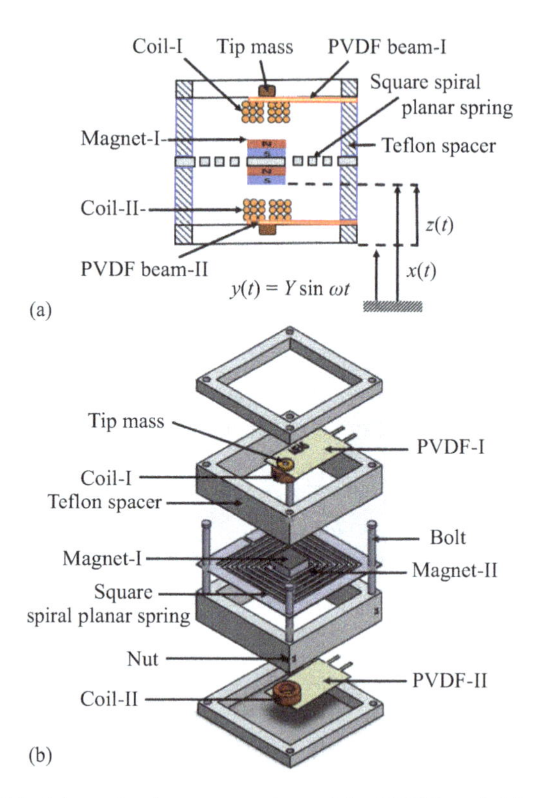


*Figure 8.1 HIEH: (a) vertical cross-section of the HIEH and (b) an exploded view of the HIEH*

proof mass, both on top and bottom sides, right at the center of the spring. The spring follows a square spiral pattern, in order to increase the length of the beam and thus, results in lowering the resonant frequency to the minimum possible value, while attaining the walking step frequency. Nuts and bolts (2 mm) in Teflon spacers are used to fix the spring firmly in the device. The upper and lower flexible PVDF cantilever beams are firmly clamped to the spacers, with wound coils attached to their tips, such that the wound coils are just in line and positioned as close to the magnetic masses to ensure a maximum flux density over the coils. The tip mass of the PVDF cantilever (coil and brass) acts as a driving force and lowers down the resonant frequencies of the beams. Both the beams are provided with individual supporting spacers to add gaps between the coils and magnets.

The PVDF beams are tuned at different resonant frequencies to the square spiral planar spring, so as to ensure multiresonant states and wide operation frequency of the harvester.

The working principle of the developed HIEH is based on PE–EM coupling, using EM and PE transductions. When the harvester experiences an external excitation, the intermediate spiral planar spring starts oscillating due to the proof magnetic mass, which changes the magnetic flux density across the wound coils on the upper and lower flexible PVDF cantilever beams. As a result, an EMF is induced according to Faraday's law of electromagnetic induction. At the same time, the upper and lower PVDF cantilever beams also experience base accelerations, which produce voltages based on the PE effect (dipoles alignment due to beam's deformation and the induced voltage in the PVDF material) and deliver peak power values at resonant frequencies of the upper and lower cantilevers. The PE part of the harvester can be considered as a current source because of the large internal resistance and hence, has low power, while the EM part as a voltage source with a small internal resistance [311].

### *8.2.1 Finite element modeling*

The proposed HIEH is a resonant-type multimode system with peak output power at different resonant frequency modes as shown in the Eigen frequency (resonant frequency) analysis conducted using COMSOL Multiphysics® software, as shown in Figure 8.2. The suspended square planar spring was designed to hold disk magnets at the center, and fixed constraints were applied to the spring at the outer edges. The magnets were united with the spring, and the model was meshed to compute for the Eigen frequencies. The resonant frequencies of the HIEH are obtained at 9.6 Hz (first mode), 41 Hz (second mode), 51 Hz (third mode), and 55 Hz (fourth mode), respectively. At the first mode, the central platform of the spring holding the magnetic masses and the adjacent solid rings were moving at a maximum amplitude as shown in Figure 8.2(a). At 41 Hz, the fourth and fifth rings in the square planar ring moved at a maximum displacement, Figure 8.2(b). The displacement shifted towards the corner of the spring at higher frequencies of 50 and 55 Hz, as depicted in Figure 8.2(c) and (d), respectively. However, the upper and lower PVDF cantilever beams holding wound coils resonate at 16.5 and 25 Hz, respectively, and produced peak outputs across the upper and lower coils.

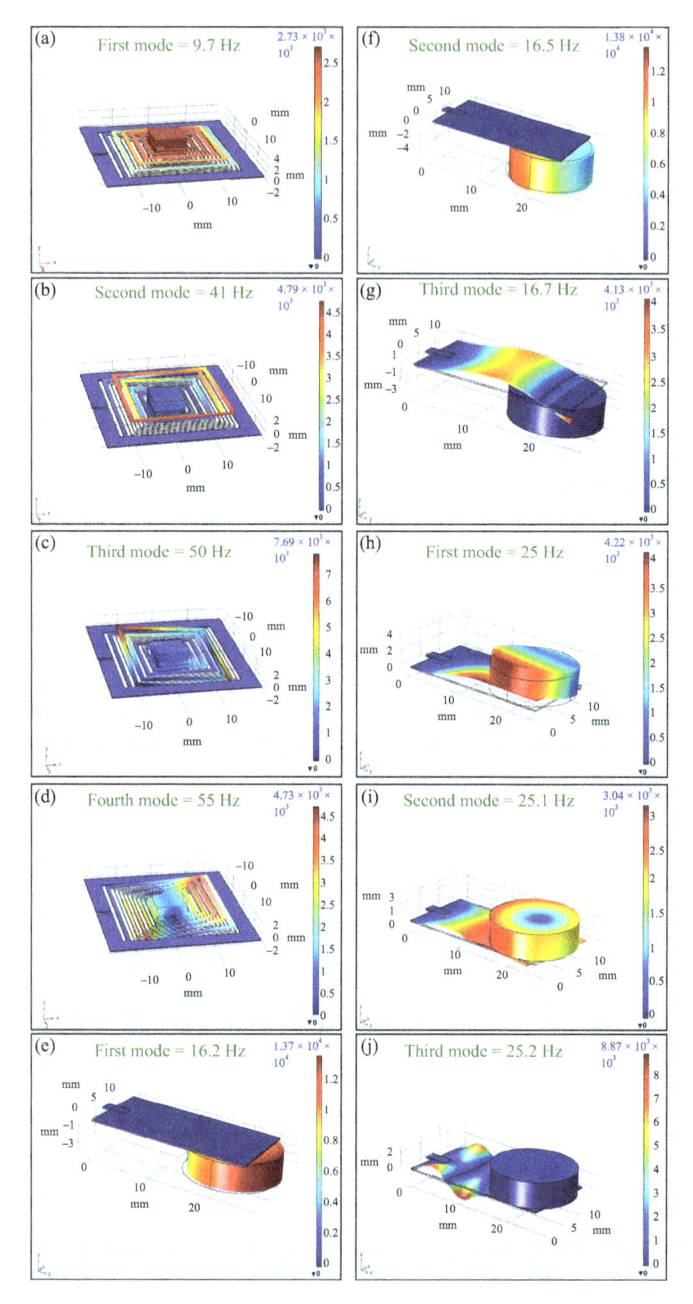


*Figure 8.2 Eigen frequency analysis of the spiral spring holding magnets: (a) first resonance at 9.7 Hz, (b) second resonance at 41 Hz, (c) third resonance at 50 Hz, (d) fourth mode at 55 Hz, (e) first mode of PVDF-I at 16.2 Hz, (f) second mode of PVDF-I at 16.5 Hz, (g) third mode of PVDF-I at 16.7 Hz, (h) first mode of PVDF-II at 25 Hz, (i) second mode of PVDF-II at 25.1 Hz, and (j) third mode of PVDF-II at 25.2 Hz*

## 8.3 Fabrication and the experimental setup

A centimeter-scale hybrid PEM-IEH fabricated in this work is shown in Figure 8.3. An intermediate square spiral planar spring of dimensions 38 mm × 38 mm × 0.26 mm was fabricated from galvanized iron (GI) steel (Shanghai Metal Co., China) using computer numerical controlled, wire-cut electrical discharge machining (CNC-EDM). The fabricated spiral spring consists of five turns, having a 1 mm wire width and a 1 mm gap between the individual spring turns. Furthermore, a platform (8 mm × 8 mm) is provided with the inner turn of the spring for magnets. Two magnets of sizes 8 mm × 8 mm × 2 mm are fixed (self-clamped) on the top and bottom sides of the central platform of the square spiral spring, as shown in Figure 8.3(b). The spiral shape of the spring has kept the size of the device to a minimum, considering the size constraints of the insole, while allowing extension of the beam length to a maximum length of 409 mm, to achieve resonance at a low walking frequency. The intermediate square spiral spring is fixed on its sides, between Teflon spacers with the same dimensions as the spring, fixing all its sides, while allowing its center to oscillate freely on exposure to external excitations, as shown in Figure 8.3(c).

An upper and lower PVDF (meas-spec) polymers (25 mm × 13 mm) were used as cantilever beams. Two wounds of conducting coils (Ø 12 mm × 4 mm), coil-I and coil-II, were produced from a 80 μm enameled copper wire, and fixed to the under side (to face magnets), at the tip of cantilever beams, and just in-line and close to the respective (upper and lower) magnets for maximum power generation.

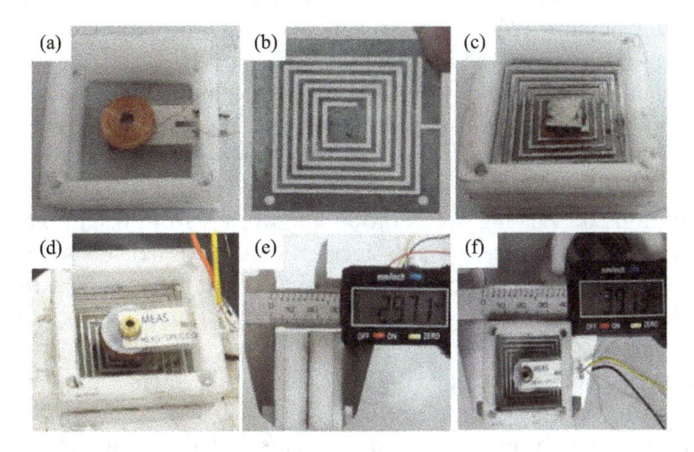


*Figure 8.3 Photographs of the developed HIEH during the assembly stages: (a) PVDF-I holding wound coil-I is clamped between Teflon spacers, (b) square spiral spring holding with a central platform for the top and bottom magnets, (c) the spring holding top and bottom magnets are sandwiched between the spacers, (d) a top view of the assembled HIEH, (e) height of the harvester, and (f) side length of the HEH*

*Table 8.3 Geometric features of the developed HIEH*

| Feature | Dimensions |
|---|---|
| Width of each turn of the spiral spring | 38 mm |
| Thickness of the spiral spring | 0.27 mm |
| Length of the spiral spring | 409 mm |
| Young's modulus of the spring material (GI steel) | 200 GPa |
| Width of PVDF beam-I and II | 13 mm |
| Thickness of PVDF beam-I and II | 0.153 mm |
| Length of PVDF beam-I and II | 25 mm |
| Tip mass on PVDF beam-I and II | 0.72 g |
| Coil-I and II size | Ø 12 mm × 4 mm |
| No. of turns in coil-I | 430 |
| Coil-I resistance | 13.5 Ω |
| No. of turns in coil-II | 470 |
| Coil-II resistance | 16.5 |
| Dimensions of the magnet | 8 mm × 8 mm × 2 mm |
| Mass of each magnet | 1.24 g |
| Gap between the coils and magnets | 3 mm |
| Harvester's overall dimensions | 39.1 mm × 39.1 mm × 29.7 mm |

PVDF-I and PVDF-II were securely clamped by the upper and lower Teflon spacers, with its terminals left outside to allow easy electrical connections and measurement of output signals, as shown in Figure 8.3(d).

The harvester was assembled using small 2 mm nuts and bolts. The developed HIEH comprises an upper hybrid generator (PVDF-I and magnet-coil-I) and a lower hybrid generator (PVDF-II and magnet-coil-II). The geometric parameters of the developed HIEH are listed in Table 8.3.

The output performance of the HIEH was tested inside the laboratory using the experimental setup: a schematic diagram is depicted in Figure 8.4(a), and a developed experimental setup is shown in Figure 8.4(b). The harvester was firmly fixed on the vibration shaker's table, with a vibration shaker used to generate sinusoidal input excitations of varying acceleration amplitudes from 0.1 to 0.6 g and frequencies from 1 to 115 Hz. A 3-axis accelerometer (Model: EVALADXL335Z, Norwood, MA, USA) was attached to the shaker's table to record the different base acceleration (g) levels of input frequency signals generated from the function generator (Model: GFG 8020H, GW Instek, New Taipei, Taiwan). A 12 V, DC power supply (Model: GT-41132, GlobTek, Inc., Japan) was used to supply power to the power amplifier (Model RM–AT2900, Rock Mars, United Arab Emirates), which magnifies and regulates signals to the vibration shaker, to excite the harvester. Moreover, a digital oscilloscope (Model: GDS-2204A, GW Instek, New Taipei, Taiwan) and digital multimeters (Model: UT81A/B, Uni-Trend Technology, China) were used to measure and analyze output signals from the accelerometer and the harvester, respectively.

In the upper and lower hybrid generators, both the PE and EM outputs were integrated. Both outputs are in-phase and are connected in series resulting in

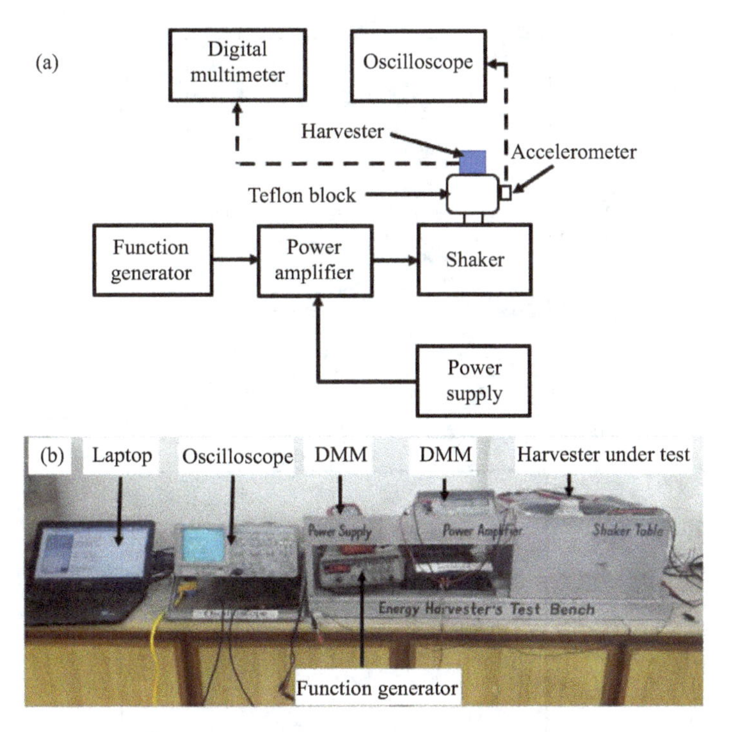


*Figure 8.4 (a) Schematic of the experimental rig and (b) the developed experimental setup for in-lab characterization of the HIEH*

improved output voltage, which is nearly equal to the sum of the individual output voltages. To obtain the optimum power for the HIEH, the output voltage was measured across different external load resistances.

## 8.4 Experimental results

The HIEH was characterized inside the laboratory under sinusoidal input excitation for a frequency sweep (1–115 Hz) of varying acceleration amplitudes from 0.1 to 0.6 g. Figure 8.5(a) and (b) shows the output voltage of the upper and lower hybrid generators for varying frequencies at 0.1, 0.4, and 0.6 g acceleration levels. The device operates in a wide operating frequency of 45 Hz, exhibits multiresonant states corresponding to the resonant frequencies of the upper and lower cantilever beams, and an intermediate square spiral spring under FFS. The spiral spring responds to low-frequency oscillations of 5 Hz and resonates at 9.7 Hz, producing the highest no-load voltage of 1.41 V across a lower hybrid generator under 0.6 g base acceleration, as illustrated in Figure 8.5(a). In this resonance mode (9.7 Hz), the spring central platform and magnets move up and down relative to the coil. Moreover, the square spiral spring excites at 41, 50, and 55 Hz, produces the second and third peak voltages of 4.55 and 1.8 V, respectively, at the second and third

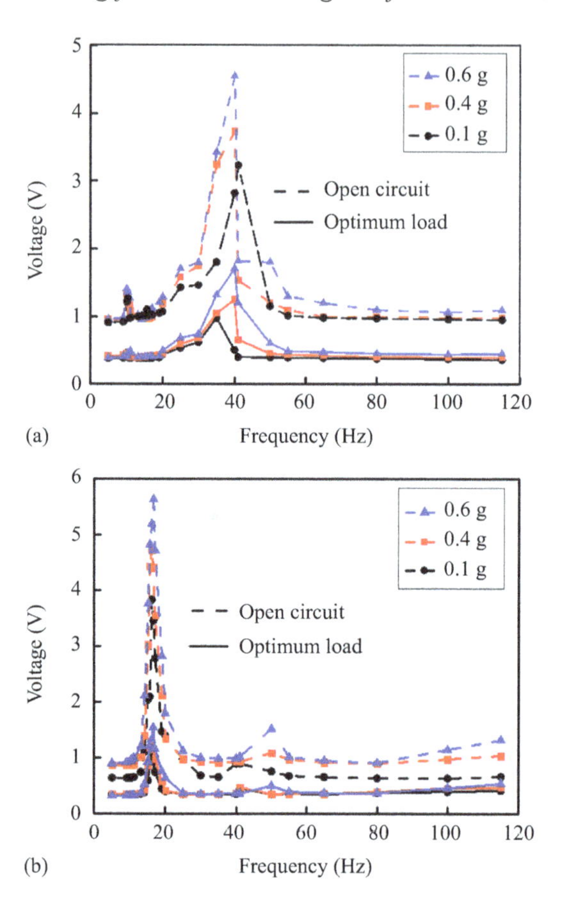


*Figure 8.5 HIEH subjected to different acceleration levels: (a) frequency response of the lower hybrid generator for no-load and optimum load resistance and (b) frequency response of the upper hybrid generator across no-load and optimum load resistance*

modes of vibration. Furthermore, the lower PVDF cantilever resonates at 25 Hz and produces an open-circuit voltage of 1.71 V under 0.6 g base acceleration across the lower hybrid harvester.

The upper PVDF cantilever excites at 16.5 Hz and delivers an open-circuit voltage of 5.64 V across the upper hybrid generator under 0.6 g, as depicted in Figure 8.5(b). The upper and lower hybrid generators are also tested across matching impedances (optimum loads) at different acceleration levels during FFS as represented by the solid lines in Figure 8.5(a) and (b). The peak voltage levels of 1.55 and 1.71 V under 0.6 g are delivered to the optimum resistances.

Figure 8.6(a) and (b) shows the variation of the RMS voltage and average power levels as a function of different load resistances between 5 and 400 Ω connected across coil-I and coil-II, respectively. The HIEH is kept oscillating at the

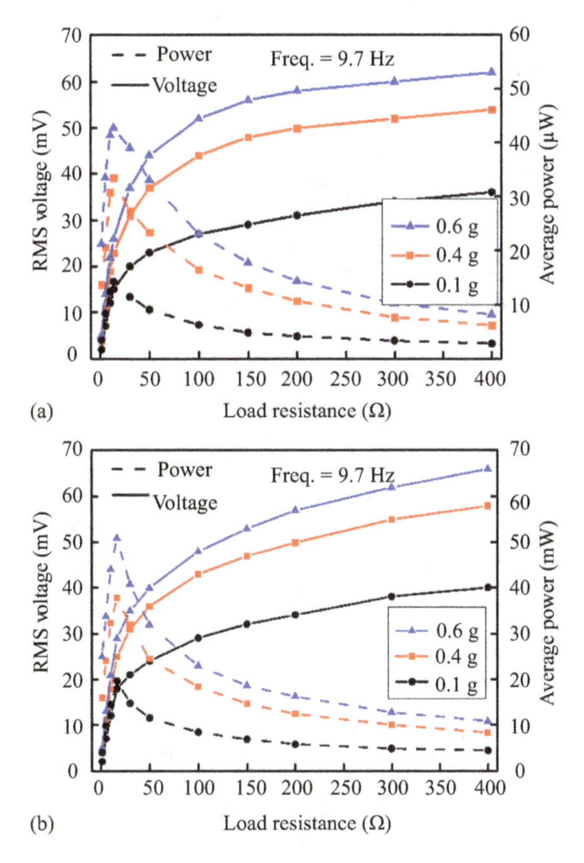


*Figure 8.6 Dependence of load voltage and power of the coil-I on the external load resistance at resonance (9.7 Hz) under 0.1, 0.4, and 0.6 g base acceleration and (b) load voltage and power as a function of various load resistances across coil-II under 0.1, 0.4, and 0.6 g base acceleration at 9.7 Hz*

first resonant frequency of 9.7 Hz and subjected to different base accelerations. The behavior of upper and lower EM generators is identical but with peak voltages at different base accelerations as depicted in Figure 8.6(a) and (b). With increasing load across the upper and lower EM generators, the output RMS voltage initially increases considerably before becoming gradually flatter at relatively higher loads. Average power, however, increases until it reaches the highest value at an optimum load (13.5 Ω for coil-I and 16.5 Ω for coil-II) before decreasing exponentially. With average power obtained from the RMS voltage [247], using equation $P = V_{\mathrm{RMS}}^2/R_L$, maximum powers are delivered across both the coils with load resistances equal to the coil's internal resistance, which satisfies a maximum power transfer [229].

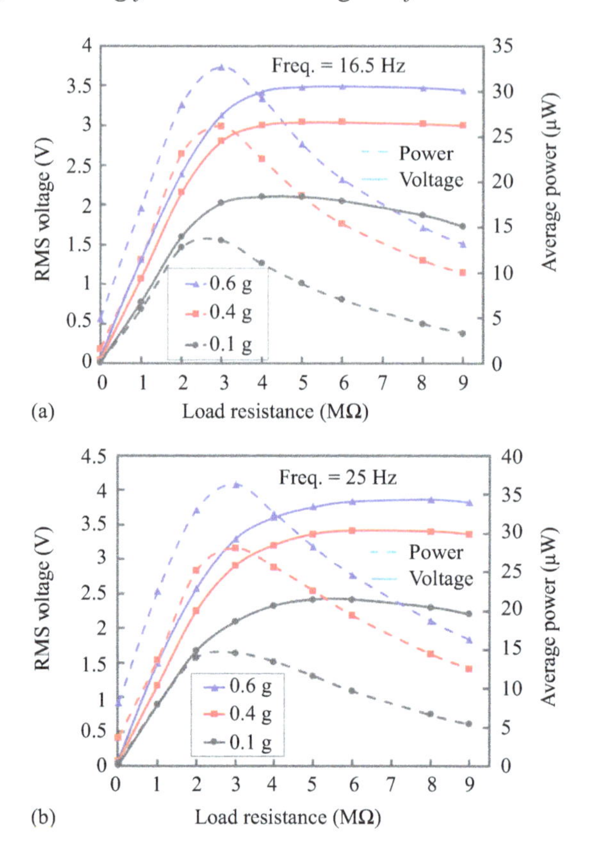


*Figure 8.7 (a) RMS voltage and average power versus load resistance across PVDF-I at 16.5 Hz under different base accelerations and (b) RMS voltage and average power versus load resistance across PVDF-II at 25 Hz under different base accelerations*

Figure 8.7(a) and (b) indicates the dependence of load voltage and load power on load resistance values across the PVDF terminals. Under 3 MΩ load resistance, PVDF-I and PVDF-II show the highest average power when excited at resonant frequencies of 16.5 and 25 Hz, respectively. The RMS voltage and an average power increase with increasing amplitude of base acceleration, as depicted in Figure 8.7(a), and the highest load voltage of 3.4 V were obtained across PVDF-I at 16.5 Hz under 0.6 g acceleration at a loading resistance of 9 MΩ.

As demonstrated in Figure 8.7(b), the RMS voltage and average power increase with an increase in base acceleration and the peak power reaches 36.3 μW across a 3 MΩ load resistance at the third resonant frequency of 25 Hz of the lower cantilever beam. The power produced by the HIEH was demonstrated by charging a 100 μF capacitor and on successful integration into the sole of a commercial shoe. The harvester was connected to a full-wave rectifier

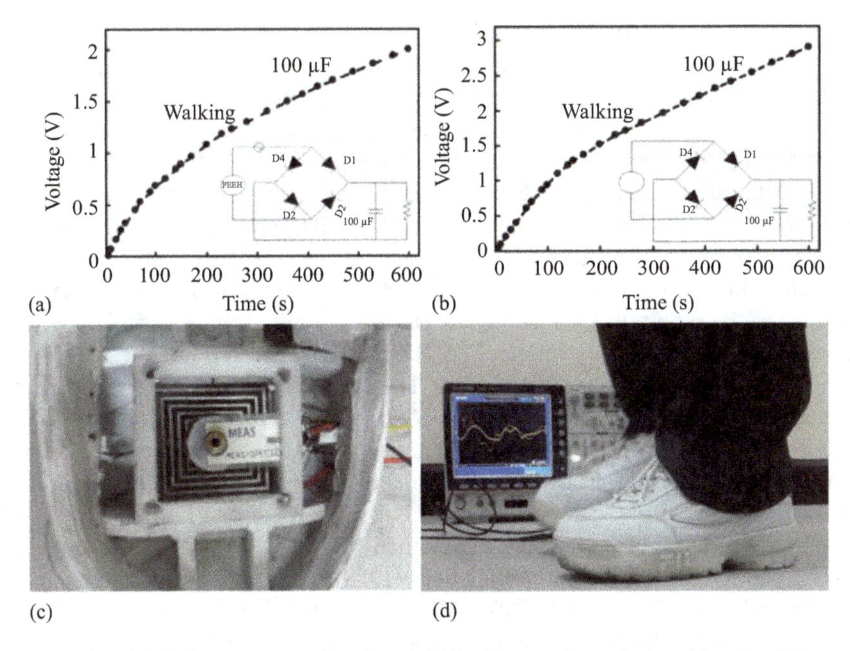


*Figure 8.8 (a) Voltage curve showing a 100 μF capacitor charged by the PE part of the harvester, (b) a capacitor charged by the hybrid PEM-IEH, (c) photograph of the HIEH incorporated into the sole of a commercial shoe, and (d) voltage generation is shown in the oscilloscope with footstep fall*

and is used to charge a 100 μF capacitor from 0 to ~1.8 V with normal walking for 8 min from the PE portion, as shown in Figure 8.8(a). The same capacitor was charged up to 2.9 V by the hybrid PE–EM coupling in same time, as shown in Figure 8.8(b). The hybrid harvester when integrated into a commercial shoe, as shown in Figure 8.8(c), maintained a stable voltage supply from 5 to 50 Hz and a better-charging performance than that of the individual EM or PE unit, as shown in Figure 8.8(d).

## 8.5 Comparison and discussion

Most of the IEHs reported in the literature are either PE, triboelectric, EM, or hybrids by combining two or three of the aforementioned harvesting techniques. The developed HIEH is compared with previously developed harvesters based on important parameters such as the resonant frequency, acceleration level, device's internal resistance, voltage and power generation capabilities, device size, and the operation mechanism. PE transduction and triboelectrification are used by most of the reported IEHs with a few utilizing EM inductions. The electrostatic mechanism, however, has been rarely used in the insole because of the initial charge requirement in these harvesters [312]. Different PE polymers such as PVDF and PE ceramics, such as lead zirconate titanate (PZT) and aluminum nitride (AlN), are commonly used in insole with multiple beam geometries. Triboelectric materials with different patterns (curved, parallel, and zigzag) have also been increasingly utilized in insole applications due to their lower resonant frequencies, strong electronegativity, cost-effectiveness, robust, and simple integration. In EMIEHs, the coil is extensively made of copper (Cu) wire because of good conductivity, ductility, and tensile strength while aluminum (Al) is used for the suspension unit owing to its good flexibility, nonpermanent deformation, and good fatigue strength. Combining two or more harvesting mechanisms in a hybrid system is a recent research interest for the sustainable drive of microelectronics. The resonant frequencies of the reported IEHs, shown in Table 8.4, range from 3 to 50 Hz, with the highest operating frequency range of 45 Hz, and base accelerations to which these harvesters were subjected were 0.1–1.0 g. The reported triboelectric IEHs generally generated more voltage levels (75–134 V) than those of PEEHs (20–30 V) and EMIEHs (0.22–0.24 V). However, the internal impedances of the PEEHs (400 kΩ–2 MΩ) and TEEHs (15–120 MΩ) are generally more than those of the EMEHs (12–240 Ω). Therefore, triboelectric (1.67–84.7 μW) and PE (30.55–800 μW) generators produce relatively less power than that of EM (61.3–1 150 μW) generators. The powers generated by the HIEHs (109–32 000 μW) are more than those of the standalone PEEHs (4.9–800 μW), EMEHs (61.3–1 150 μW), and TEEHs (1.67–11 700 μW) due to the combined transduction mechanisms.

The HIEH developed in this work is a low resonant-type multimodal system being able to operate at a wide operation frequency (9–55 Hz) and is compared with the reported PEEHs, EMIEHs, TEEHs, and HIEHs, as listed in Table 8.4. Having dual transduction mechanisms, multiresonant states, compact size, lightweight, comparatively lower internal impedance, and the ability to operate at low-frequency vibrations makes it a power-efficient system among the reported IEHs.

## 8.6 Summary

A HIEH based on PE–EM transduction has been developed and tested. The HIEH is able to power wearable electronics by scavenging biomechanical energy during walking. The harvester which constitutes an upper and lower hybrid PE–EM generator is based on the effective conjunction of piezoelectricity and EM induction and generated an overall peak power of 109 μW and 70 μW, corresponding to power densities of 2.47 and 1.58 μW/cm$^3$, respectively. Keeping into consideration the available space inside the shoe sole and the walking step frequency, the devised HIEH is designed as a compact structure and exhibited six resonant states in a lower frequency range resulting in a wider operation frequency of approximately 45 Hz. Furthermore, the hybrid generator has a better charging performance than that of the standalone unit and can supply power sustainably to wearable gadgets, like, a pedometer, a smartwatch, and wireless body-monitoring sensors. The intermediate spring is holding magnetic masses on the top and bottom sides and the magnets are inline and close to the wound coils attached to the PVDF beams. On subjecting to the input sinusoidal signals on the shaker's table, first, the square spiral spring starts oscillating at a frequency as low as

*Table 8.4 Literature summary of standalone and HIEHs*

| Insole harvester's type | Harvesting mechanism | Internal impedance (Ω) | Resonant frequency (Hz) | Base acceleration (g) | Open-circuit voltage (V) | Device size ($cm^3$) | Peak power (μW) | Power density (μW/$cm^3$) | Power density per acceleration (μW/g/$cm^3$) | Ref. |
|---|---|---|---|---|---|---|---|---|---|---|
| Standalone | PE | 400k | — | — | 30 | 16.8 | 800 | 47.61 | — | [124] |
| | | 150k | 5.6 | 2 | 7 | 5 | 43 | 8.6 | 4.3 | [225] |
| | | 3.3 M | 3 | — | — | 576 | 5 | — | — | [247] |
| | | 2 M | 12 | 0.55 | 20 | 2.56 | 30.55 | 11.91 | 21.65 | |
| | EM | 12 | 9 | 0.8 | 0.24 | 47.1 | 1 150 | 24.41 | 30.5 | [2] |
| | | 5 | 9.1 | 0.85 | | 78.5 | 420 | 5.35 | 6.29 | [138] |
| | | 800 | 8 | 0.04 | 134 | 53.38 | 14.55 | 2.85 | 71.25 | [230] |
| | | 240 | 1 | — | 0.22 | 20.1 | 61.3 | 3.04 | — | [234] |
| Hybrid | Triboelectric–EM–PE | 1 M, 70, | — | — | 75 | 50 | 32 000, 33 000 | 660 | — | [158] |
| | EM-triboelectric | 6 M, 2k | — | — | 268, 5 | 62.5 | 4 900, 3 500 | 78.4, 56 | — | [159] |
| | PE–triboelectric | 10 M, 32 M | 25 | 1 | 186 | 12.5 | 774 | 61.92 | 61.92 | [303] |
| | Triboelectric-EM | — | — | — | 13.2 | 10.5 | 39 864 | 3 796.5 | — | [309] |
| | Triboelectric-EM | 6 M, 1.5 | 2 | — | 15 | 1.84 | 29.8, 16.7 | 16.19, 9.09 | — | [313] |
| | PE–EM | 13.5, 16.5, 3 M, and 3 M | 9.7, 16.5, 25 41, and 50 | 0.1–0.6 | 7.01 | 44.1 | 179 | 4.05 | 6.76 | [160] |

5 Hz and reaches its first resonance at 9.7 Hz. The upper and lower PVDF beams resonate at 16.5 and 25 Hz, producing a peak power of 33 μW and 37 μW across stretchable PVDF cantilevers under 0.6 g acceleration at matching the impedance of 3 MΩ. The energy was stored using a 100 μF capacitor through walking, and as compared to the PE unit, the hybrid harvester charged the same capacitor at 30% more voltage at the same time as compared to that of the standalone PEH. The charging capacity of the HIEH shows that stored energy can be used to operate health-monitoring body sensors and microelectronic gadgets.

*Chapter 9*

# Overview of the finite element analysis and its applications in kinetic energy harvesting devices

## 9.1 Introduction

Modern electronic and portable consumer devices, such as wearable devices, sensor nodes, IoT devices, etc., require an external power source. In general, these low-power devices are powered by electrochemical batteries (e.g., NiCd, NiMH, Li-Po, Li-ions, etc.) that require periodic battery replacement or charging for the devices to perform. Frequent battery replacement or charging can be a hassle for remotely planted devices such as sensor devices. To overcome these problems, a self-powered energy-harvesting device can be integrated to sustain the device operation [314–316]. These problems also provide opportunities for an alternative energy source to power up these devices and prolong their lifespan. Renewable energy that is available and can be harvested within those device environments such as light, thermal, and kinetic energy. Abundant kinetic energy available around us from the movements [317–320], vibrations [321–323], and motions of different objects [324–327] has gained a lot of interest for researchers to develop small-scale KEH devices. The KEHs convert the usually wasted kinetic energy into the form of movements into usable electrical energy. Different types of KEH systems, namely, PE, EM, triboelectric, and hybrid EM–TE systems, were discussed in much detail in the previous chapters.

A PEEH converts kinetic energy in the form of vibrations or mechanical force applied into electrical energy through the PE effect. Mechanical movements can be in form of linear or nonlinear movements depending on the device setup. An EMEH design typically consists of springs, magnets, and coils. Mechanical movements cause a change in the magnetic flux of the systems which produces an EMF force. The difference between those two is that PE typically works with high-frequency range kinetic energy, while the EM device works in a low-frequency range. A TEEH harvests energy by utilizing contact electrification, i.e., sliding and contact or separation between two TE materials. TE materials used by researchers and developers include polymers, metals, and inorganic materials. Commonly used materials for THE are dielectric materials such as polytetrafluoroethylene (PTFE), fluorinated ethylene propylene (FEP), polydimethylsiloxane (PDMS), and kapton [328]. Among these energy-harvesting devices, PE-based devices are the most favorable ones for researchers because of their simple design and the ease of setup compared to the others.

To study the initial design stages and for developing these small-scale energy harvesters on the prototype level, researchers usually utilize computational methods, i.e., FEA to simulate the devices with different input parameters such as dimensions, materials, and material properties, and other different working parameters, etc. The FEA is an advanced tool for predicting and solving numerical problems across many fields of physics and engineering during the designing stage. It is used to analytically predict the behavior or the reaction of certain elements under specified conditions, i.e., vibrations, motion, friction, pressure, etc. The FEA is an approach that allows the researchers to virtually verify their design before final fabrication and experimental testing. It also allows us for predicting the potential failure and behavior of the dependent/independent physical systems or devices in a close resemblance. In general, the FEA workflow consists of following steps:

Step 1: Problem definition
Clearly define the problem to be analyzed using FEA. This may involve specifying the type of analysis (e.g., structural, thermal, fluid, etc.), identifying the physical system or structure to be analyzed, and defining the boundary conditions and loads applied to the system.
Step 2: Preprocessing
Generate a geometric model of the system using CAD software. This involves creating a digital representation of the physical structure or system to be analyzed.

Define the finite element mesh, which is a discretized representation of the geometry. The mesh consists of small, interconnected elements that approximate the shape of the physical structure.

Assign material properties and boundary conditions to the finite elements. This includes specifying the material properties, such as the Young's modulus, thermal conductivity, or fluid viscosity, as well as applying boundary conditions such as constraints, loads, and initial conditions.
Step 3: Formulation
Formulate the governing equations that describe the behavior of the system based on the finite element model and the problem definition. This typically involves applying the principles of mechanics, heat transfer, or fluid dynamics to create a set of partial differential equations that govern the behavior of the system.

Define the type of analysis to be performed, such as static, dynamic, or thermal analysis. This determines the specific equations and solution techniques to be used.
Step 4: Assembly
Assemble the finite element equations into a system of equations that can be solved numerically. This involves combining the element-level equations to obtain a global system of equations that represents the behavior of the entire system.

Assemble the stiffness matrix, which represents the relationship between the applied loads and the displacements or temperatures at each degree of freedom in the finite element model.

Incorporate any additional considerations such as contact, nonlinear material behavior, or other special conditions into the formulation.
Step 5: Solution
Solve the system of equations using numerical methods to obtain the unknown values of the variables in the model. This may involve techniques such as direct or iterative solvers, time-stepping methods for dynamic analysis, or numerical integration for transient analysis.
Monitor the convergence of the numerical solution to ensure the accuracy and reliability. Adjust the solution parameters if necessary.
Extract the results of interest, such as displacements, stresses, temperatures, or fluid velocities, from the numerical solution for further analysis.
Step 6: Post-processing
Analyze and interpret the results obtained from the numerical solution. This may involve visualizing the results using plots, graphs, or animations to gain insights into the behavior of the system.
Evaluate the accuracy and reliability of the results by comparing them with experimental data, analytical solutions, or design specifications.
Make engineering decisions or recommendations based on the analysis results, such as design optimizations, performance improvements, or safety evaluations.
Step 7: Iteration
If necessary, iterate the analysis by refining the finite element mesh, adjusting the boundary conditions, or modifying the model based on the results and feedback from the post-processing step.
Repeat the preprocessing, formulation, assembly, solution, and post-processing steps as needed to obtain more accurate and reliable results or to investigate different scenarios or design alternatives.
Step 8: Post-processing
Draw conclusions from the analysis results and use them to make informed decisions for design optimization, performance improvement, or further analysis iterations.
Document the analysis process, results, and conclusions for future reference and communication with stakeholders.
It is important to note that the specific steps and details of the FEA process may vary depending on the type of analysis being performed, the FEA software being used, and the problem being solved. It is crucial to follow the guidelines and documentation provided with the respective software.

This chapter reviews the use of FEA applications for KEH devices, especially on how FEA is applied to simulate and predict the outcome of the KEH systems. The utilization of FEA makes early concept decision-making and optimizing the model design easier, and significantly reduces the need for fabricating physical prototypes.

## 9.2 FEA applications for KEH devices

PE devices are one of the most studied devices due to their simple design. Researchers have been constantly trying to optimize and improve the PEH design.

To reduce the number of errors and testing phases, researchers implement the FEA methods in their study for predicting the output of the system such as power output, voltage output, shear and stress, deformation and displacement, resonant frequency, etc.

In a study carried out by Wang *et al.* [329], FEA was used as a validation tool for their numerical calculation for a unimorph PEH, where the analysis of the effect of structural damping and electromechanical coupling was carried out to optimize power and vibration amplitude. Figure 9.1 shows the FEA model used in their study, in which a resistor element is connected to the PE element. The PEH cantilever is subjected to vibrations in a vertical direction with a given amplitude and angular frequency.

Zhang *et al.* [330] proposed a linear-arc composite beam, and a detailed finite element simulation is used to virtually simulate the acceleration of the mass and resonant frequency in their design. The FEA was also used to determine the best curvature design for the KEH design. Figure 9.2(a) shows the mesh division at the region of interest. It is worth noting that the less relevant section (end mass) was kept sparse compared to the beam section. This is used to optimize the solver calculation thereby reducing the simulation error. Figure 9.2(b) and (c) shows the stress cloud and voltage cloud of the linear arc-beam KEH. It can be seen that the stress is higher near the base of the system. The stress cloud was used to virtually predict the amount of stress in the beam. In their design, a higher stress value will give a higher output voltage. Based on the FEA results, it was predicted that 40 $m^{-1}$ arc curvature produces the highest open-circuit output voltage, which is astonishingly 44.3% higher than that of a straight beam design.

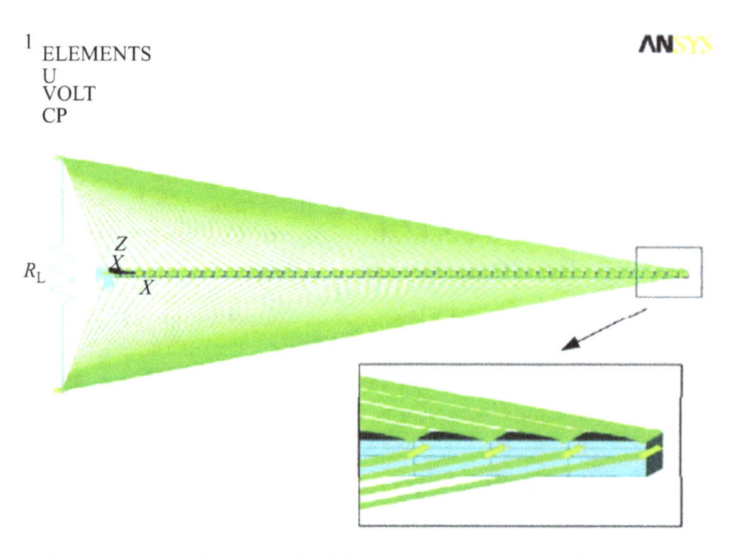


*Figure 9.1 FEA model of the PEH connected with a resistor*

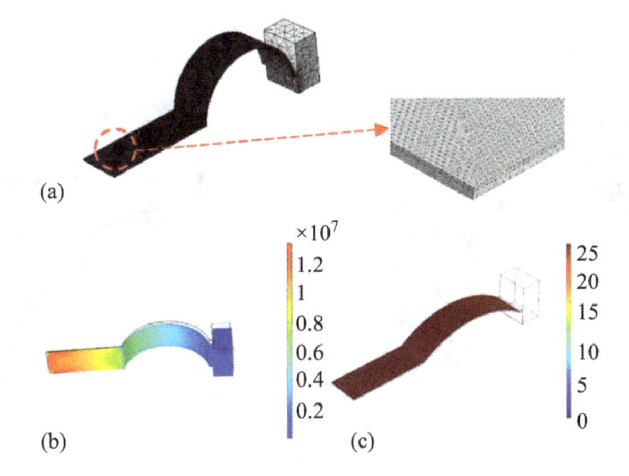


*Figure 9.2 Linear-arc composite beam proposed by Zhang et al. (a) mesh division, (b) stress output cloud, and (c) voltage output cloud*

Xie *et al.* [331] designed an E-shaped PEH with built-in bistability and internal resonance located in the middle of the beam. The built-in bistability is used to eliminate the use of an external magnet for fabricating a compact structure device. The design consists of a U-shaped beam and a parasitic cantilever beam (inner middle section). FEA simulation studies were used to obtain the optimized vibration modes and to analyze the performance of the design under different excitation conditions (Figure 9.3(a)). The FEA was successful in calculating the first-order vibration at 8.51 Hz, and also estimated a deformation displacement localized near the U-shaped beam, while the second-order vibration at 16.58 Hz shows a displacement of the parasitic cantilever beam, thus estimating both performance and weak points in the structure. These overlapping displacements can enhance the frequency bandwidth of the system. Upadrashta and Yang [332] proposed a finite element model of the nonlinear broadband 2-degree-of-freedom energy harvester (Figure 9.3(b)). Here, FEA is used to simplify the magnetic force interaction model of the PEH system and predict the estimated output voltage with a minimal base acceleration. A common PEH device application ambient vibration is found to be of a value of $<3$ m/s$^2$, the proposed PEH design is predicted to be able to harvest energy at a base acceleration of 2.83 m/s$^2$.

Ju and Ji [321] demonstrated the use of FEA analysis to verify the analytical model of their system and the deformation of the PE cantilever beam in their impact-based PE device. The deformation or vibration is caused by the motion of the tungsten carbide element installed inside the energy harvesting system. A natural frequency of vibration at 9 997 Hz was used as a virtual motion of the human body. Figure 9.4(a) illustrates an impact-based system using a tungsten carbide ball (spherical proof mass) as a source of vibration for the PE cantilever beam. The results help us to determine the minimal force required and thickness of the beam before failure. Gao *et al.* [333] have developed a wearable insole PEH device.

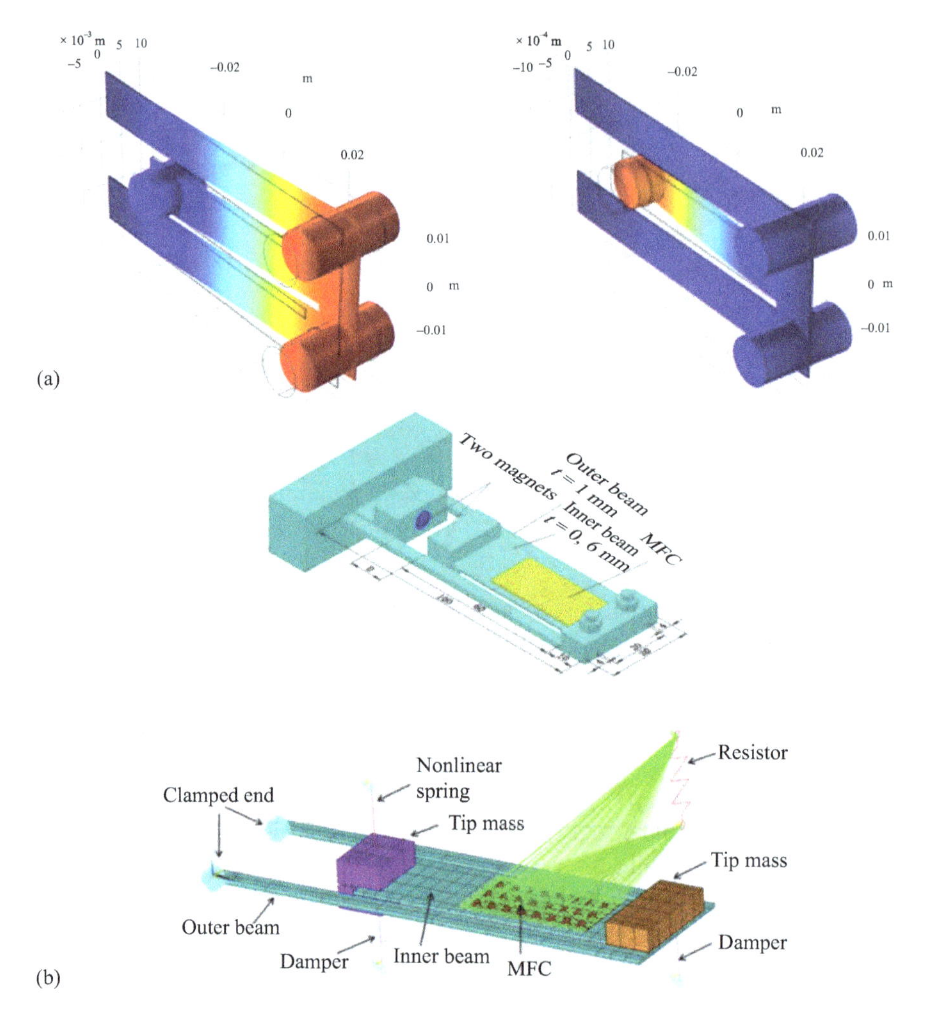


*Figure 9.3 E-shaped PEH: (a) first-order vibration of 8.51 Hz frequency (left) and second-order vibration of 16.58 Hz (right) by Xie et al. and (b) a schematic diagram (top) and the finite element model (bottom) proposed by Upadrashta and Yang*

In order to optimize the wearable PEH unit design, FEA analysis was performed based on several factors such as anisotropic PE membrane and isotropic shell and frames, rigid connection between all members, and compression load on the top plate (Figure 9.4(b)). A walking frequency was applied at the top plate from 1 to 2.2 Hz with a step of 0.2 Hz to study the voltage output of the PEH system.

Other examples of FEA studies on PEH devices are asphalt pavement embedded with PEH [334], regenerative shock absorber PEH [317], and rotary-type PEH [320, 335]. The tire-pavement PEH proposed by Du *et al.* [334] as shown in Figure 9.5(a)

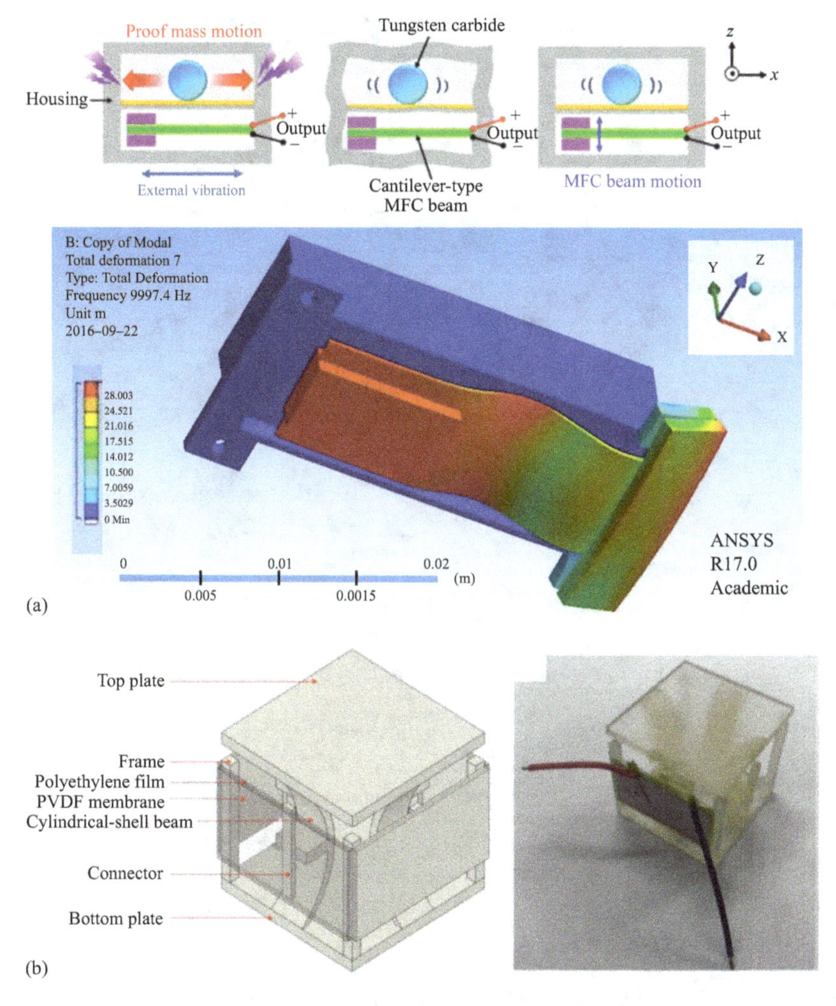


*Figure 9.4 Impact-based PEH: (a) a schematic diagram (top), and (b) deformation analysis results revealed by Ju and Ji*

shows the multiscale PEH transducer behavior under an applied load. The FEA is used to illustrate the real-world behavior of the system for optimizing the system, carload, and speed of a car. The FEA study indirectly helps engineers to reduce the damage to the pavements embedded with PEH devices. Figure 9.5(b) illustrates the prototype of the regenerative shock absorber to transfer the kinetic energy into usable PE energy. The design consists of a bevel-gear mechanism and a rack-and-pinion mechanism. The FEA is used to evaluate the static structural analysis of the assembly design. The design was subjected to various boundary conditions and resonating forces, i.e., load, bending, torsion, and stress, to obtain the natural frequencies at which the assembly will resonate. A rotary-type PEH is shown in Figure 9.5(c), as proposed by Noralia *et al.* [335]. The upper surface PEH patch is subjected to a sinusoidal working

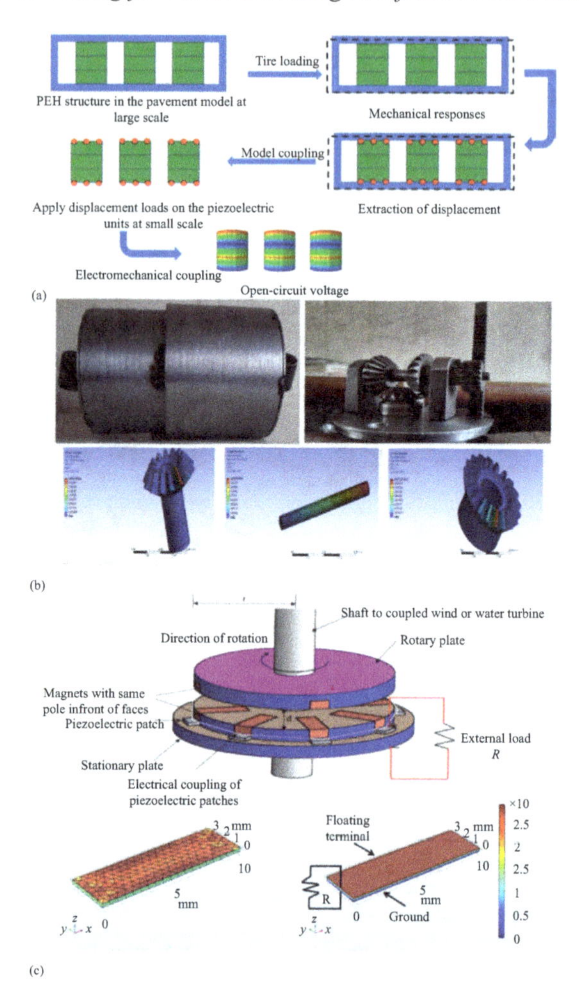


*Figure 9.5 FEA studies of PEH transducer devices (a) asphalt pavement design proposed by Du et al., (b) a regenerative PEH shock absorber proposed by Dahat et al., and (c) a rotary-type energy harvester proposed by Narolia et al.*

force generated from the monopolar magnets that are attached to the upper rotary plate of the shaft. An initial FEA analysis was conducted by Narolia *et al.* [320] to study the shear mode of the PEH system for rational motion. This FEA study analyzes the effect of PE elements with different parameters (length, width, and thickness), magnetic block spacing, angular speed, and external load for optimization of the final design. In a later study [335], FEA was used to analyze the minimum torque required to rotate the rotary plate for experimental purposes.

An innovative design proposed by Farhangdoust *et al.* [314] is a self-powered high-efficiency PEH device for a health-monitoring device. The device design is

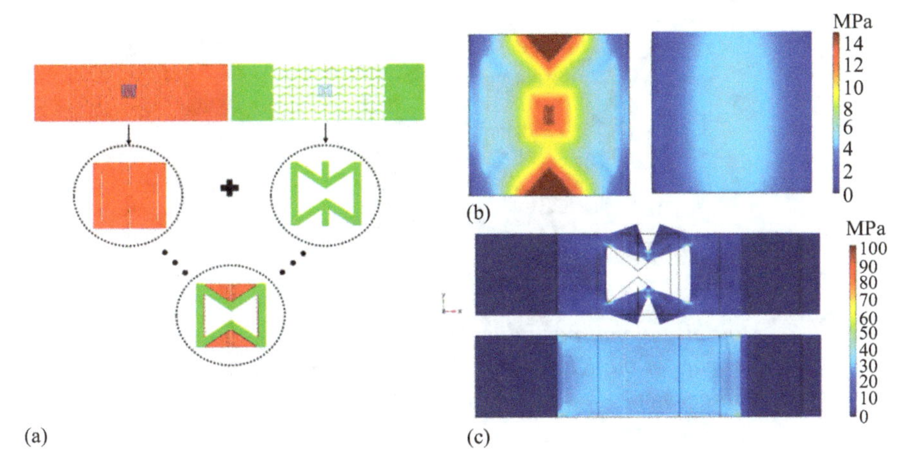


*Figure 9.6 Kirigami and auxetic topology substrate design for mounting PE element proposed by Farhangdoust et al.: (a) MetaSub design Kirigami (red) and auxetic (green) topologies, (b) PE element stress map distribution, and (c) displacement of the MetaSub*

based on a combination of Kirigami and auxetic topologies metamaterial-based substrate (MetaSub). The MetaSub is used to mount a PE material that is subjected to a periodic stretch and compression as shown in Figure 9.6. FEA studies were carried out to investigate the performance of the PEH under a sinusoidal strain cycle and to prove the feasibility of the design at low- and high-frequency simulation tests. Another innovative self-powered PEH device such as using hybrid composite materials [326] and a pendulum-based PEH device [316] have been stressfully achieved. Ham *et al.* [326] developed a self-powered hybrid composite ceramic-polymer hybrid lead-free PE particles and P(VDF-TfFE) matrix for sensing kinetic movements. FEA simulations were used to confirm the effectiveness of the matrix for PE potential output by employing the physical variables of the materials and calculating mechanical strain. Zhang *et al.* [316] developed a multidirectional pendulum KEH based on homopolar repulsion. In this design, a homopolar magnet was placed on a PE beam and the pendulum swing to provide a repulsion effect (Figure 9.7). The FEA simulations were carried out to investigate the maximum displacement and stress region caused by the pendulum swing at various swing angles (from 0° to 70°) and pendulum mass with a weight of 10–40 g.

An EMEH device typically consists of a coil-and-magnet setup. Magnetic induction (i.e., Faraday's law of electromagnetic induction) from the movement magnet and coil creates changes in magnetic flux to generate current or energy. The area of interest of the EMH devices includes the coil setup, a spring platform, and a magnet arrangement to produce a high output at a minimal vibration.

Utilization of FEA in optimization of a linear EMH design has been demonstrated by different researchers. Zeng *et al.* [324] introduced linear motion KEH as shown in Figure 9.8(a), which consists of a series of coils and a

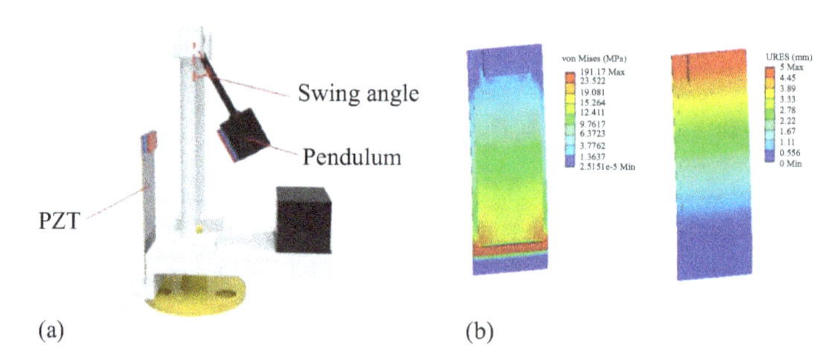


Figure 9.7 Multidirectional pendulum swing PEH devices: (a) the schematic diagram of the KEH device and (b) the maximum stress and displacement FEA results at $F = 0.623$ N

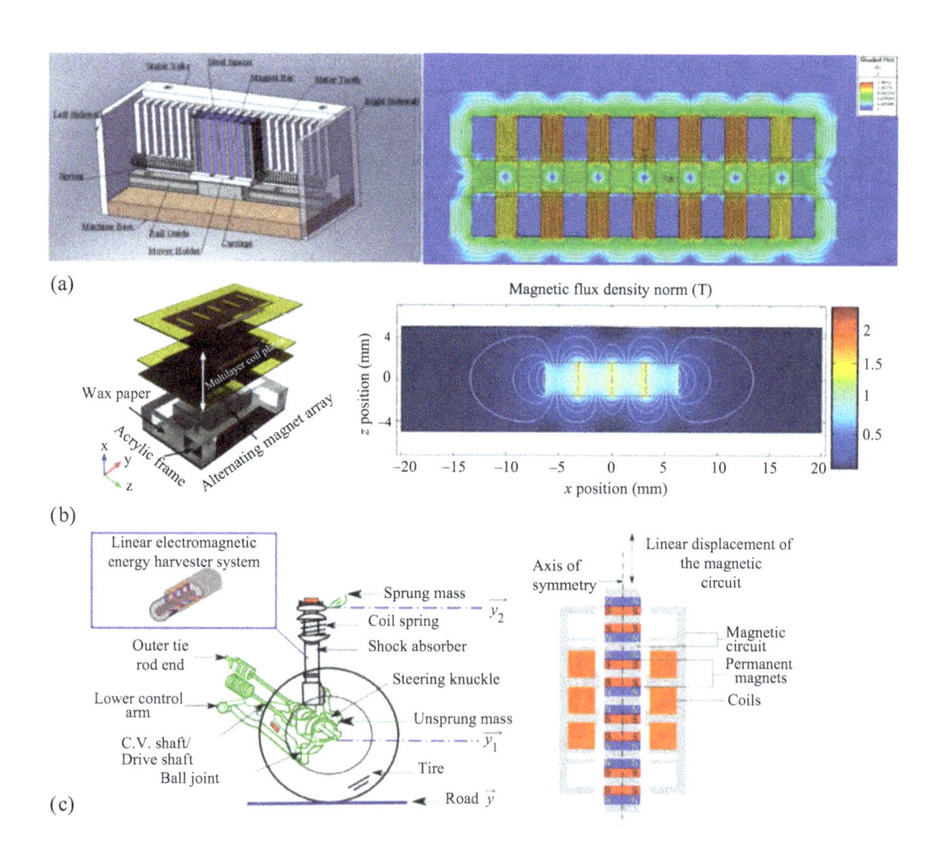


Figure 9.8 A linear EMH system design: (a) coil optimization study by Zeng et al., (b) microfabricated coils EMH designed by Fan et al., and (c) suspension absorber EMH designed by Lafarge et al.

series of magnets attached to a spring and FEA helped us immensely in the optimization of the number of coil turns. Fan *et al.* [336] utilized FEA to propose the design of their microfabricated coil sheet for a nonresonant broadband EMH device (Figure 9.8(b)). Magnetic flux density was studied to illustrate the excitation motion effect on the system. An EMH system embedded on a vehicle suspension by Lafarge *et al.* [318] was used to harvest energy from the linear motion of the suspension. The FEA simulation method was used to derive the flux evolution of magnetic induction. The magnetic flux results are used to determine the magnetic forces generated by the coils and the tooth of the magnetic circuit (Figure 9.8(c)).

Halim *et al.* [325] utilized the FEA method for their EMH to predict the power generated from pseudo-walking excitation that mimics the swing motion of human-arm during walking or running. Figure 9.9(a) shows the arrangement of the magnets used for FEA simulation for their device. FEA simulations were used to predict the rate of change in magnetic flux w.r.t. to the relative angular displacement of the pole-pairs as the magnets move past the coil. This allows them to study the best possible arrangement, strength, and size of the magnets and coils required for the system. A novel biaxial pendulum-based vibration energy harvester has been developed by Luo *et al.* [319], where a hemispherical pendulum is presented to have the capability to simultaneously alternate bidirectionally to adapt the path of vibration excitation. EM analysis was carried out using FEA to obtain the spatial magnetic field distribution (Figure 9.9(b)) to predict the relationship between vibration excitation and the output voltage of the model. Paul *et al.* [315] designed a tapered nonlinear vibration EMH device. In this design, a flexible tapered spring structure was used as a platform that is able to vibrate to create a magnetic induction effect. The FEA was used to study the nonlinear movements to optimize the displacement and stiffness of the tapered spring.

A hybrid KEH can be used to fully utilize and capture abundant kinetic energy generated from the surrounding motions. Hybrid KEH systems were developed by Shi *et al.* [322] and Gupta *et al.* [337]. Shi *et al.* [322] designed a floating PE and EM hybrid KEH for harvesting wave vibration energy actuated by a rotating wobble ball. In this setup, PE materials and driving magnets are attached vertically to the cantilever beam. The rotation of the wobble ball creates a magnetic induction current over the copper coil, and the deformation of the cantilever beam creates a PE effect. A separate FEA study is carried out to analyze the performance of PE transducers and EM generators (Figure 9.10(a)). In a hybrid KEH mechanism developed by Gupta *et al.* [337], an EMH is used to harvest the kinetic energy from the vibration of the PDMS spring and TEH to harvest the separation of the two tribo-materials (ITO and PTFE). Figure 9.10(b) shows a schematic diagram of the hybrid device and the FEA simulation study to predict the displacement of the spring at a natural frequency of 58.561 Hz (top) and 108.55 Hz (bottom).

Table 9.1 shows the compilation of various types of KEH devices produced by researchers with the help of FEA tools.

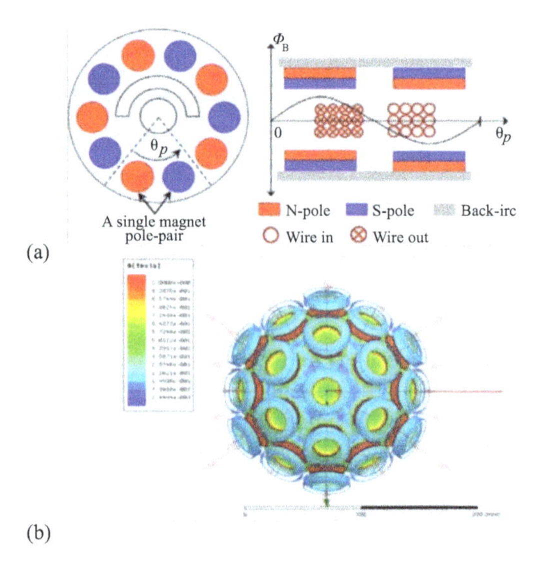


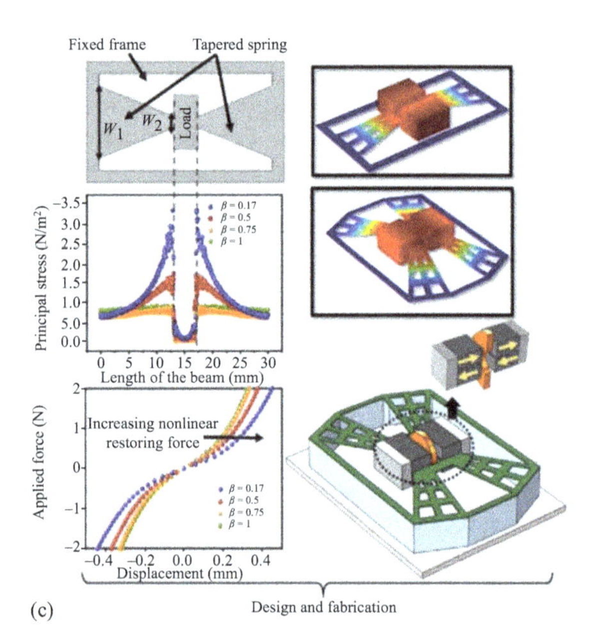


*Figure 9.9 A novel nonlinear design for EMH system: (a) an EMH generator developed by Halim et al. for hand movement motion, (b) a hemispherical pendulum EMH developed by Luo et al., and (c) a tapered spring EMH design developed by Paul et al.*

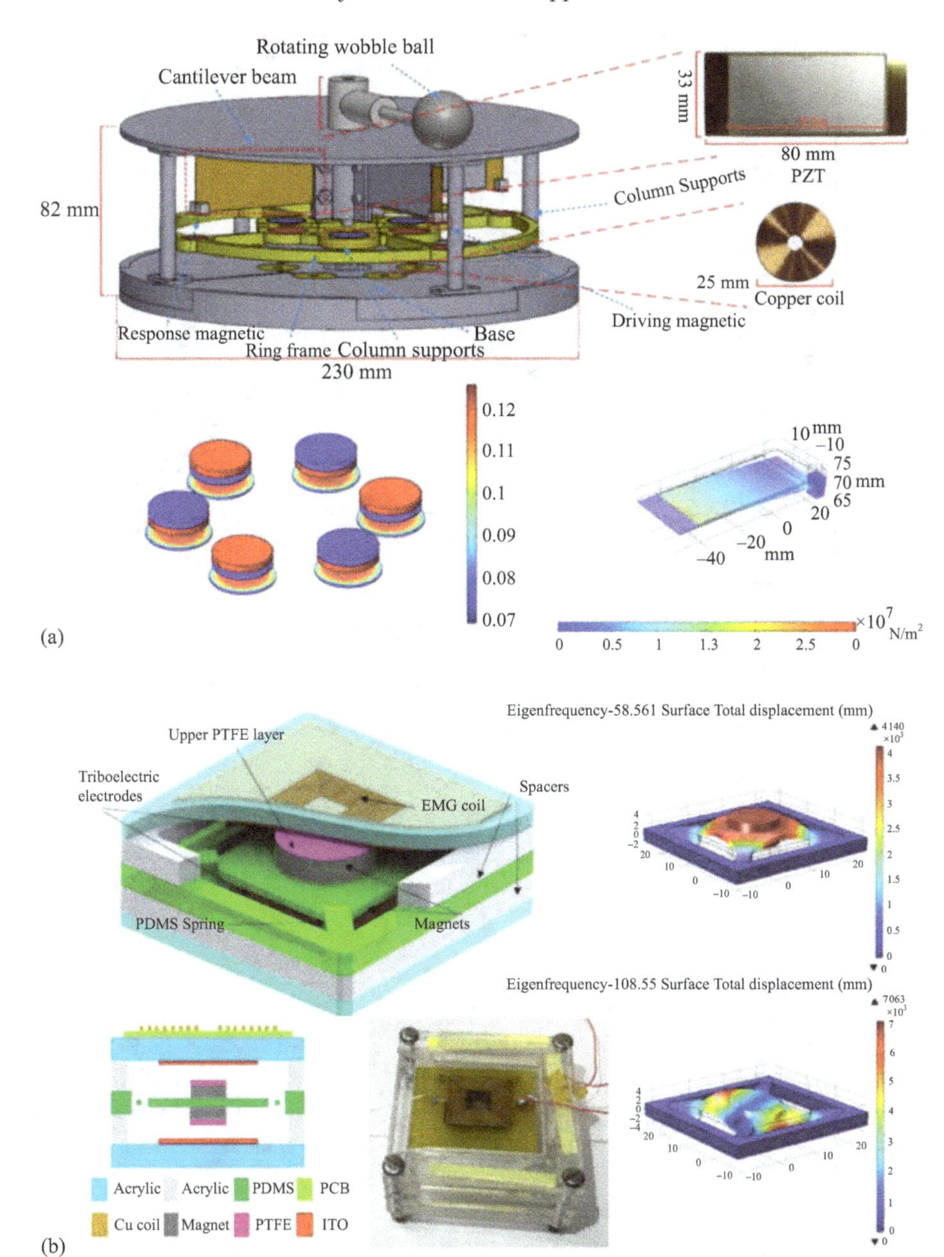


*Figure 9.10 A hybrid KEH systems: (a) PE/magnetic hybrid system proposed by Shi et al. and (b) an EM/TE hybrid mechanism proposed by Gupta et al.*

*Table 9.1 KEHs using the FEA analysis to validate experimental results*

| Type | FEA simulation | | | Experiment (E) Numerical (N) | | | Ref. |
|---|---|---|---|---|---|---|---|
| | Frequency (Hz) | Voltage (V) | Power ($\mu$W) | Frequency (Hz) | Voltage (V) | Power ($\mu$W) | |
| PE | 48.2 ($V_{sc}$) 49.2 ($V_{oc}$) | — | — | — | 47.8 ($V_{sc}$) 48.8 ($V_{oc}$) | — | [329] |
| PE | 15 | 25 | — | 14.5 | 24 | — | [330] |
| PE | 7 and 18 | 50 | — | 6 to 18.3 | 56.1 | 786.8 | [331] |
| PE | 19.4 | ~48 (rms) | — | 19.5 | ~49 (rms) | — | [332] |
| PE | 9 997 (Resonance frequency) | — | — | 17 | 42.2 (P2P) | 633.7 | [321] |
| PE | 2.2 (Single PVDF layer) | 1.43 (Single PVDF layer) | 11 (Single PVDF layer) | — | 4.15 (Triple PVDF layer) | 8 600 (Triple PVDF layer) | [333] |
| PE | — | 1 500 ($V_{oc}$) @ axle load 25 N | — | — | — | — | [334] |
| PE | — | — | — | 6 (Displacement 15 mm) | 47.7 | | [317] |
| PE | — | — | — | 140 (rpm 2 100 rpm) | 0.04 | 0.01448 | [335] |
| PE | — | — | — | 600 rpm | — | 358.644 W | [320] |
| PE | 10 | — | 165 | — | — | — | [314] |
| PE | | | | | | | [326] |
| PE | — | — | — | — | 13.6 (40 g mass, 70° swing) | 1233 | [316] |
| PE | — | — | — | — | 5 ($V_{oc}$) (90° bending) | ~1.2 | [326] |
| EM | — | 0.886 (Peak); 0.444 (rms) [165 turns coil winding] | 434.5 | — | 1.08 (Peak); 0.431 (rms) [165 turns coil winding] | 497.7 | [324] |

*(Continues)*

| | | | | | | |
|---|---|---|---|---|---|---|
| EM | — | — | — | 8 | — | 346.7 | [336] |
| EM | — | 20.1 | 7.2 | — | 20 | 7 | [318] |
| EM | — | — | — | 1 (±25 rotational amplitude) | — | 61.3 | [325] |
| EM | — | — | — | 1.5 to 2.2 | 14.25 | 2.03 W | [319] |
| EM | — | 2.5 ($V_{oc}$) (2g acceleration) | — | — | 2.2 ($V_{oc}$) (2 g acceleration) | 550 (1.6 kΩ load resistance) | [315] |
| Hybrid (PEH- EMH) | — | — | — | 1.4 | — | PEH = 21 950<br>EMH = 11 260 | [322] |
| Hybrid (PEH-TEH) | | | | 80 (Excited frequency)<br>68 (Bandwidth range) | | 50.2<br>Power density = 0.8 $\mu W/cm^3$ | [337] |

## 9.3 Applications and future directions

The FEA method is a sophisticated and widely used simulation tool that is highly regarded by engineers and product designers for its versatility and reliability in predicting the behavior of complex systems. In the context of KEH devices, FEA plays a pivotal role in providing valuable insights into the deformation and stress of PE materials, magnetic flux in EM devices, and other critical parameters that impact the device performance.

FEA enables engineers and researchers to accurately model and analyze the behavior of KEH devices without the need for destructive testing, which can be time-consuming and expensive. Through FEA simulations, engineers can virtually test different design iterations, evaluate the performance under varying conditions, and optimize the device performance before physical prototypes are built. This not only saves time and resources but also allows for a deeper understanding of the device behavior, leading to more informed design decisions.

Commercial FEA software packages, such as ANSYS, COMSOL, SimScale, Autodesk CFD, and others, have huge material libraries and testing environments for preprocessing, processing, problem-solving, and post-processing tasks. These software packages are equipped with advanced solvers and algorithms that can accurately capture the complex physics involved in KEH systems, including PE effects, magnetic fields, and fluid dynamics, among others. This enables engineers to perform comprehensive analyses and obtain detailed results for different operating conditions, leading to a more robust and optimized device design.

Furthermore, FEA has also revolutionized the analysis of microscale-sized devices, which are gaining increasing attention due to their potential for high throughput and miniaturization. FEA allows for the virtual analysis of complex microscale structures in both 2D and 3D environments, providing valuable insights into the behavior of these devices at a detailed level that was previously unattainable. This has opened new opportunities for researchers and engineers to design and optimize microscale KEH devices with improved efficiency and performance.

The increasing awareness and utilization of FEA simulation tools in the field of KEH offer significant potential for the development of compact and efficient devices in the future. By leveraging the power of FEA, researchers and industries can gain valuable insights into the behavior of KEH systems, leading to more innovative and sustainable designs. The predictive capabilities of FEA in analyzing complex systems, its ability to model microscale devices, and its extensive material libraries and testing environments make it an indispensable tool for engineers and researchers in the pursuit of advancing KEH technologies [314,336–340].

# *Chapter 10*
# Energy harvesters for biomechanical applications

## 10.1 Introduction

The field of biomechanics has witnessed remarkable progress in recent years, thanks to the revolutionary breakthroughs in self-powered technology and smart materials. These cutting-edge developments have opened up exciting new avenues for the creation of highly innovative implantable energy harvesting and biomechanical applications. With such advancements, the possibilities for the future of biomechanics are truly limitless [341]. Biomechanical energy harvesting is a breakthrough technology that harnesses the power of natural body movements to generate sustainable energy. Using PE materials, this technology can efficiently convert mechanical energy into electrical energy, opening up new possibilities for implantable devices, sensors, and other biomedical applications. By utilizing the body's own mechanics and movements, biomechanical energy harvesting offers a reliable and ecofriendly alternative to conventional power sources. Its potential to transform the field of biomechanics is immense, as it enables the development of more advanced and innovative devices that can enhance the quality of life for people worldwide [342].

PE materials, capable of producing an electrical output in response to stress or strain, are among the promising materials in this field. The PEEH technology has great potential to benefit various biomechanical devices, including health monitoring, cell stimulation, brain stimulation, and tissue engineering. By generating sustainable electrical energy from mechanical energy, PEHs can offer a reliable and efficient alternative power source for implantable devices, sensors, and other biomechanical applications, as shown in Figure 10.1.

In the context of wearable and mobile electronic devices, biomechanical energy harvesting is a technology that can potentially revolutionize the way these devices are powered. By utilizing the body's natural movements, such as walking, running, or even typing, biomechanical energy harvesting can generate sustainable energy that can be used to power these devices. This means that users can potentially have longer battery life and reduced dependence on external power sources [344]. With the help of PE materials, this technology can efficiently convert mechanical energy into electrical energy, making it a reliable and eco-friendly alternative to traditional power sources. Biomechanical energy harvesting has the

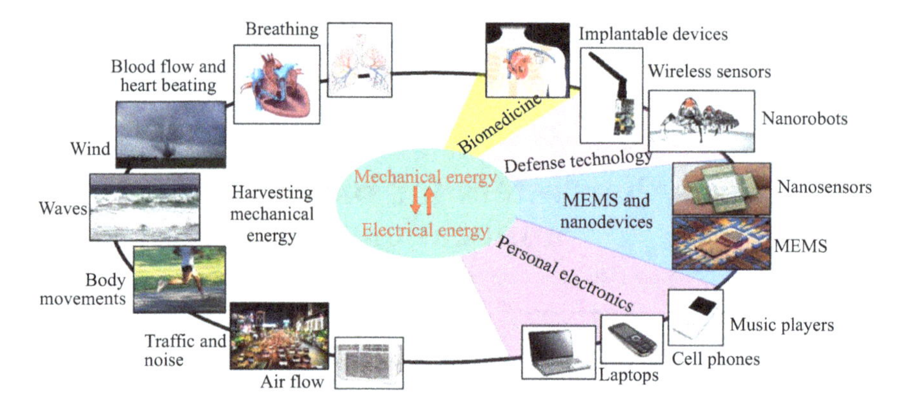


*Figure 10.1 Energy harvesting [343]*

potential to significantly enhance the performance and sustainability of wearable and mobile electronic devices [345,346].

The potential of energy-harvesting technologies for biomechanical applications, particularly in the field of wearable and mobile electronic devices, is significant. PEEH technology utilizing PE materials has shown promise in converting mechanical energy into electrical energy for various biomedical applications, including health monitoring, cell and brain stimulation, and tissue engineering. Additionally, energy-harvesting technologies, including biomechanical energy harvesting, can help IoT devices improve their performance, reduce energy consumption, and prolong the battery life. By leveraging the body's natural movements and other sources such as temperature gradients, vibration, radiofrequency signals, and solar light, energy-harvesting technologies can lead the way to more sustainable and efficient devices in the future.

## 10.2 Biomechanical energy

Biomechanical energy refers to the energy generated by the movements and mechanical actions of living organisms. It is a vast and widely used source of energy, especially in the case of humans. The human body has numerous joints and pressure points that can be utilized for energy harvesting through activities such as limb movement, respiration, chest and abdomen displacement, eyelid movement, and even body pressure exerted while walking [347]. Biomechanical energy-harvesting techniques offer significant potential for various applications, such as powering wearable and mobile electronic devices, health monitoring, cell and brain stimulation, and tissue engineering. Additionally, these techniques can help reduce energy consumption and prolong the battery life of IoT devices by utilizing the body's natural movements and other sources such as temperature gradients, vibration, radiofrequency signals, and solar light [348].

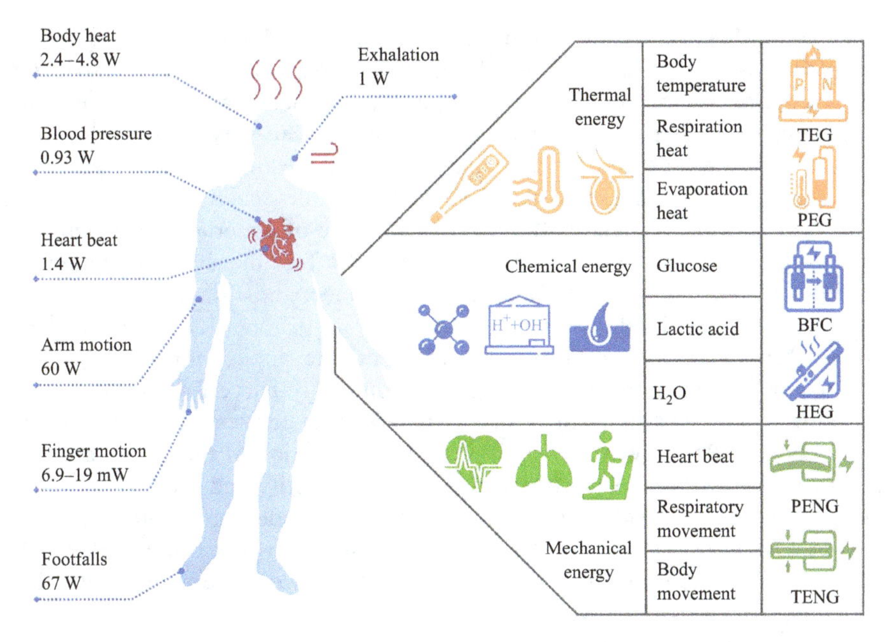


*Figure 10.2 Human body energy sources and applicable energy-harvesting technologies [349]*

The human body is a significant source of biomechanical energy that can be harvested using various technologies, as shown in Figure 10.2. The body has numerous flexible joints and areas of high-pressure concentration, which provide opportunities for energy harvesting through limb movement, respiration, chest and abdomen displacement, eyelid movement, and body pressure onto soil while walking. These sources of energy can be converted into electrical energy using a variety of energy-harvesting technologies such as PE, EM, and TE generators. Efficient harvesting and utilization of these energy sources can lead to the development of self-powered and sustainable electronic devices for various applications, including health monitoring, environmental sensing, and wireless communications.

## 10.3 Key considerations for biomechanical energy harvesting

When designing and implementing biomechanical energy-harvesting systems, there are several key considerations that must be taken into account. These include the choice of materials, the design of the energy-harvesting mechanism, and the optimization of energy conversion efficiency. Proper consideration of these factors is crucial for the successful development of efficient and reliable biomechanical energy-harvesting devices. In addition to the key considerations mentioned earlier, other factors that need to be taken into account when designing biomechanical energy-harvesting systems include the intended application and the characteristics of the excitation source. For example, the frequency and amplitude of the body movements that serve as the excitation source can vary greatly depending on the specific application. Therefore, the energy-harvesting mechanism needs to be designed to accommodate these variations while maintaining an optimal energy conversion efficiency. Moreover, the choice of materials for the energy-harvesting mechanism is also critical, as it can affect the device's durability, weight, and the overall performance. For instance, the PE material used for energy conversion should be selected based on its PE coefficients, mechanical properties, and biocompatibility, depending on the application. Finally, it is essential to evaluate the performance of an energy-harvesting system using appropriate metrics, such as the power output, energy conversion efficiency, and power density. Proper evaluation allows for optimization of the design and performance of the device, ensuring its long-term reliability and usability. Overall, taking these key considerations into account when designing biomechanical energy-harvesting systems can lead to the development of more efficient, reliable, and sustainable energy-harvesting devices for wearable and mobile electronic applications.

### *10.3.1 Excitation sources for biomechanical energy harvesting*

Applications for biomechanical energy harvesting for wearable and mobile devices rely on natural body movements as the excitation source. These movements can include walking, running, typing, or other physical activities. The excitation source needs to be carefully chosen to ensure that it is sustainable and provides enough energy to power the device. Additionally, the frequency and amplitude of the excitation source must also be taken into account to optimize the energy-harvesting process. With the advent of IoT, there is an increased demand for sustainable power sources to power the countless connected devices. The potential for harvesting power from various sources such as temperature gradients, movement vibrations, radiofrequency signals, and solar light is immense. By utilizing innovative energy-efficient technologies, IoT devices can reduce their energy consumption, prolong the battery life, and improve their performance. The possibilities for harvesting power are endless, and it is exciting to see a range of sources we can tap into to achieve a more sustainable and efficient IoT ecosystem [350]. As shown in Figure 10.3, there are a variety of technologies and methods available for harnessing these energy sources and transforming them into usable power. By taking advantage of these innovative energy-harvesting techniques, we can reduce our dependence on traditional power sources and move towards a more sustainable and environmentally friendly future.

Optimizing power consumption is a crucial aspect of enhancing the performance of smart devices. By implementing innovative energy-efficient technologies, these devices can not only reduce their energy consumption but also extend

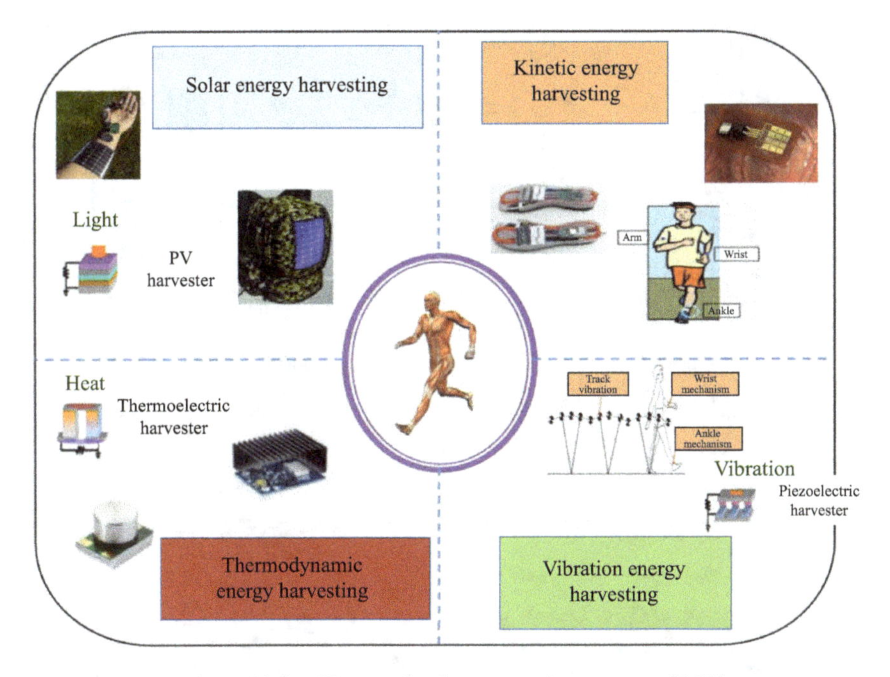


*Figure 10.3 Human body energy harvesting [351]*

their battery life. This would ultimately result in an improved performance and greater convenience for users [352]. One such innovative technology that has emerged to address this issue is cloud computing. By leveraging the power of cloud computing, smart devices can offload resource-intensive tasks to remote servers, reducing their energy consumption and extending their battery life. This approach not only improves the performance of smart devices but also enhances their convenience for users. Thus, incorporating energy-efficient technologies like cloud computing is crucial to the development of sustainable and efficient smart devices [353].

### *10.3.2 Mechanical modulation techniques and energy conversion methods for biomechanical energy harvesting*

Once an excitation source has been identified, the mechanical modulation of the energy-harvesting system must be carefully designed to maximize the energy output. This involves selecting appropriate materials, such as PE materials that can efficiently convert mechanical energy into electrical energy. The design of the system must also account for any mechanical constraints and limitations of the device. The energy conversion process involves converting the mechanical energy generated by natural body movements into usable electrical energy. PE materials

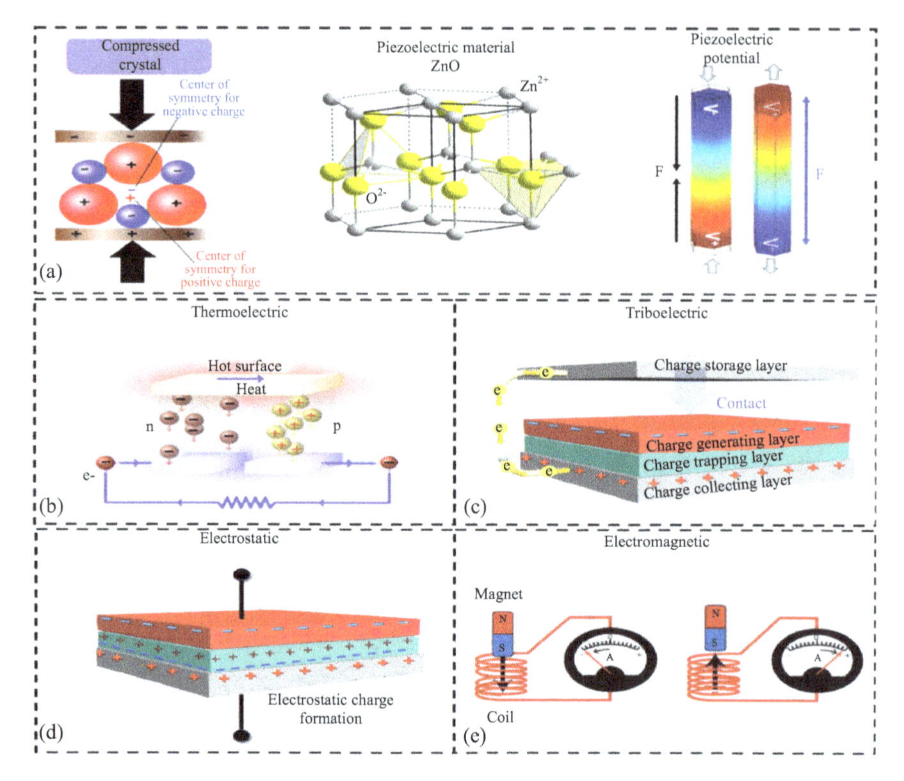


*Figure 10.4 Biomechanical energy-harvesting sources [354]*

are commonly used for this purpose, as they can generate an electrical output when subjected to stress or strain. However, the efficiency of the energy conversion process can be influenced by factors such as the size and shape of the PE element, as well as the properties of the materials used.

With the rapidly growing demand for sustainable energy sources, biomechanical energy-harvesting schemes offer a promising solution. Figure 10.4 showcases various types of energy-harvesting schemes, each with their unique working principles. The PE scheme generates electricity by compressing or straining PE crystals, while the thermoelectric scheme creates an electric potential by utilizing temperature gradients. The TE scheme generates charge through contact friction, while the electrostatic scheme stores energy through capacitive coupling. The Faraday scheme utilizes a moving magnet inside a coil to generate a potential difference. By implementing these schemes, we can effectively convert kinetic energy from human motion into electrical energy, paving the way for sustainable and efficient devices in the IoT ecosystem. It is time to embrace these innovative technologies and make a positive impact on our planet's future.

## 10.4 Evaluation metrics for biomechanical energy harvesting

The evaluation of biomechanical energy-harvesting systems involves assessing the effectiveness of the bioenergy-harvesting process in terms of the energy efficiency, output, and reliability. Evaluation metrics must be carefully chosen to accurately reflect the performance of the system. Common evaluation metrics include power output, energy conversion efficiency, and lifetime. The evaluation of these metrics must also take into account any external factors that may influence the performance of the energy-harvesting system, such as the environment in which the device is used. Evaluation metrics for biomechanical energy harvesting can be summarized in Table 10.1.

Evaluation metrics are important for assessing the performance and suitability of biomechanical energy-harvesting devices for practical applications. Power density and efficiency are critical metrics for measuring the effectiveness of the energy-harvesting mechanism in generating power from biomechanical sources. The output voltage and current are important factors for determining the compatibility of the device with the electronics it powers. Frequency response is crucial for ensuring that the device can capture the full range of biomechanical energy sources.

*Table 10.1 Evaluation metrics for energy harvesting*

| **Evaluation metrics** | **Explanation** |
|---|---|
| Power density | It is the amount of power output per unit area of the energy-harvesting device. It is measured in watts per square meter (W/m$^2$). |
| Efficiency | It is the ratio of the harvested energy output to the total energy input. It is expressed as a percentage. |
| Output voltage | It is the voltage produced by the energy-harvesting device, which is used to power the electronics. It is measured in volts (V). |
| Output current | It is the electric current produced by the energy-harvesting device, which is used to power the electronics. It is measured in amperes (A). |
| Frequency response | It is the range of frequencies that the energy-harvesting device can efficiently convert into electrical energy. It is measured in hertz (Hz). |
| Durability | It is the ability of the energy-harvesting device to withstand long-term use and harsh environmental conditions without a significant decrease in performance. |
| Scalability | It is the ability of the energy-harvesting technology to be easily scaled up for mass production and widespread adoption. |
| Cost | The total cost of materials and manufacturing for the energy-harvesting device, which impacts the affordability and accessibility of the technology. |
| Safety | The potential risks associated with the use of the energy-harvesting device, such as electrical hazards or the toxicity of materials used. It is important to ensure that the device meets safety standards and regulations. |

Durability and scalability are essential for ensuring that the device can withstand long-term use and be produced on a large scale for practical use. Cost is an important consideration for making the technology accessible and affordable for consumers. Finally, safety is crucial in ensuring that the energy-harvesting device does not pose any risks to the user.

The general equations for two commonly used evaluation metrics for bio-mechanical energy harvesting are:

(a) Power density:
Power density is a measure of the amount of power that can be harvested per unit area of the energy harvester. A common practice in the field of energy harvesting is to compare the performance of various devices. The general equation for power density is:

$$P\_\text{density} = P\_\text{output}/A$$

where *P_output* is the power output of the energy harvester and $A$ is the effective area of the harvester.

For example, if an energy harvester generates an output power of 5 mW and has an effective area of 10 $\text{cm}^2$, then the power density would be:

$$P\_\text{density} = 5\ \text{mW}/(10\ \text{cm}^2) = 0.5\ \text{mW}/\text{cm}^2$$

(b) Efficiency:
Efficiency is a measure of how well an energy harvester converts mechanical energy into electrical energy. It is defined as the ratio of the electrical power output to the mechanical power input. The general equation for efficiency is:

$$\eta = P\_\text{output}/P\_\text{input}$$

where *P_output* is the electrical power output of the energy harvester and *P_input* is the mechanical power input to the harvester.

For example, if an energy harvester converts 10% of the mechanical energy, it receives electrical energy, and the mechanical power input is 100 mW, then the electrical power output would be:

$$P\_\text{output} = 10\% \times 100\ \text{mW} = 10\ \text{mW}$$

Using this value for *P_output* and the given value for *P_input*, the efficiency would be:

$$\eta = 10\ \text{mW}/100\ \text{mW} = 0.1 \text{ or } 10\%.$$

Table 10.2 provides a comprehensive overview of the output power density, potential application areas, benefits, limitations, and different human-centric energy-harvesting techniques. It serves as a useful reference point for comparing the performance of various energy-harvesting devices.

*Table 10.2 Comparison of energy-harvesting techniques for human-centric applications*

| Energy-harvesting techniques | Power density output (mW/cm$^2$) | Characteristics | Limitations of technique | Efficiency | Potential area/applications | Ref. |
|---|---|---|---|---|---|---|
| PE | $7.3 \times 10^{-3}$ $-1.75 \times 10^{5}$ | High sensitivity; compact size; high power density; high output voltage | Output impedance is high, current is low, efficiency is minimal at low frequency, some materials are toxic, frequency-dependent performance, complex and expensive fabrication, and complicated MPPT | Moderate | Orthopedic implants, inertial sensors, pacemaker, pressure sensors, and pulse sensors | [355] |
| Ultrasonic | 1.2–290 | High efficiency for small IMDs, increased penetration depth, and compact design | Poor data transmission, low-power harvesting, requires propagation medium, and low efficiency in large IMDs | Moderate | Implantable medical devices | [355] |
| Radio frequency | 0.45–84 | Extensive variety of operating frequencies | Low output power, distance-dependent output, and unpre dictable | Low | Wireless optogenetics | [356] |
| TE | 0.0022–50 | Effortless fabrication, high density of output power, flexibility, scalability, cost-effectiveness, and the ability to use a wide range of materials, while remaining independent of frequency | High-voltage insulation, complex fabrication processes, erosion of materials, limited lifespan, and complicated interface circuitry | Low | Wearable textiles, self-powered touchpad devices, and glucose biosensors | [357] |
| Thermoelectric | 0.084–0.571 | Lightweight design, offers high reliability and accessibility, and has a long lifespan. Additionally, it is frequency-independent and offers simplified MPPT, making it easily scalable. | Low power density, low energy conversion efficiency, unpredictable harvested power, start-up circuit required, boost converter needed, and high cost | Low | ECG, pulse oximeters, pacemakers, hearing aids, EMG, and EEG devices | [358] |
| EM | $8.88 \times 10^{-3}$–5.614 | The technique has high efficiency, high power density, and a simple MPPT system, making it robust and capable of producing high-output currents | Efficiency decreases at low frequency; miniaturization leads to lower efficiency; potential source of EMI; complicated integration with MEMS; the output voltage is low | Moderate | Wearable medical devices and cardiac innovator | [359] |

## 10.5 Recent designs and applications for biomechanical energy harvesting

This section highlights cutting-edge research on PEEH which harnesses human motion to power wearable and self-powered electronics. The conversation delves into recent advancements and examines the potential of PEEH as a means of generating sustainable energy in the realm of biomechanical energy harvesting. Get ready to learn about the latest configurations and designs of PEEH and discover how this technology can revolutionize the field of energy generation.

Table 10.3 presents a comprehensive roadmap and comparison for powering wearable and mobile devices with biomechanical energy harvesting.

Table 10.3 compares the parameters of biomechanical energy harvesting in the past, present, and future. In the past, PE, electrostatic, EM, and thermoelectric techniques were commonly used, while in the present, TE, hybrid, and multimodal techniques are being researched. In the future, smart and adaptive techniques are expected to be developed. The materials used have also changed from inorganic to organic and nanomaterials in the present, and are expected to become smart and self-healing in the future. Devices have also evolved from nonwearable to wearable and implantable, and are expected to become implantable and biointegrated in the future. The energy conversion efficiency has increased from low in the past to moderate in the present and is expected to be high in the future. The applications have also expanded from military, medical, and industrial fields in the past to consumer electronics, healthcare, and IoT fields in the present and are expected to include personalized healthcare, smart cities, and environmental monitoring in the future.

*Table 10.3 Roadmap and comparison*

| Parameters | Past | Present | Future |
|---|---|---|---|
| Energy-harvesting systems | Electrostatic, EM, PE, thermoelectric | TE, hybrid, multimodal | Smart and adaptive |
| Materials | Inorganic materials | Organic materials and nanomaterials | Smart materials and self-healing materials |
| Devices | Nonwearable devices | Wearable and implantable devices | Implantable, bio-integrated devices |
| Energy conversion efficiency | Low (1%–2%) | Moderate (5%–10%) | High (>15%) |
| Applications | Military, medical, industrial | Consumer electronics, healthcare, IoT | Personalized healthcare, smart cities, environmental monitoring |

## 10.6 Biomechanical energy harvesting through smart footwear

Biomechanical energy harvesting through smart footwear has attracted significant attention in recent years due to its potential to generate sustainable energy from the natural motion of the human body. This technology involves the use of PE materials embedded in footwear to convert mechanical energy into electrical energy [360], which can then be used to power various electronic devices, as shown in Figures 10.5 and 10.6. Smart footwear has significant potential in various applications, including sports and fitness tracking, health monitoring, and rehabilitation. In this section, we discuss the latest research on smart footwear for biomechanical energy harvesting and its potential applications.

The field of biomechanical energy harvesting has witnessed significant advancements in recent years, with smart footwear emerging as a promising technology for generating sustainable energy. Smart footwear incorporates PE materials and mechanical structures to convert the energy generated during human walking or running into electrical energy, which can be used to power wearable and mobile electronic devices [362]. Latest research has focused on optimizing the design and materials of smart footwear to improve the energy conversion efficiency and device performance [363]. Various types of PE materials, including PVDF and PZT, have been tested for use in smart footwear, with PVDF showing promising results due to its high flexibility and durability [364].

In addition, researchers have explored different mechanical structures and layouts for smart footwear to enhance the amount of energy harvested [365,366]. Evaluating the performance of smart footwear has been done through various

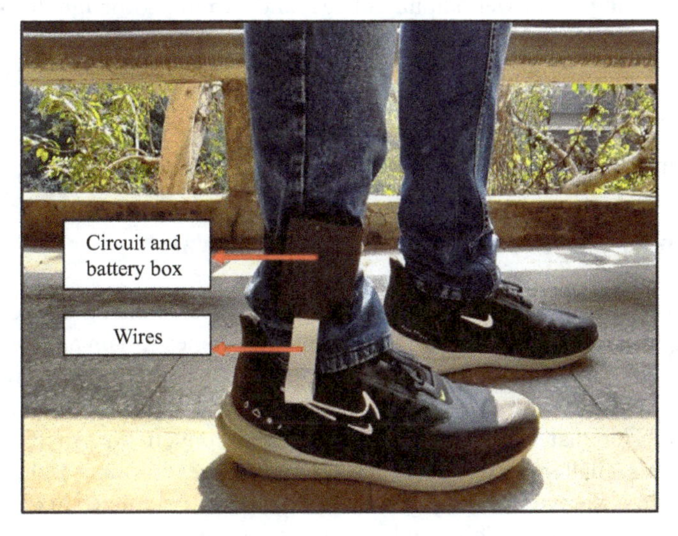


*Figure 10.5 Smart footwear [361]*

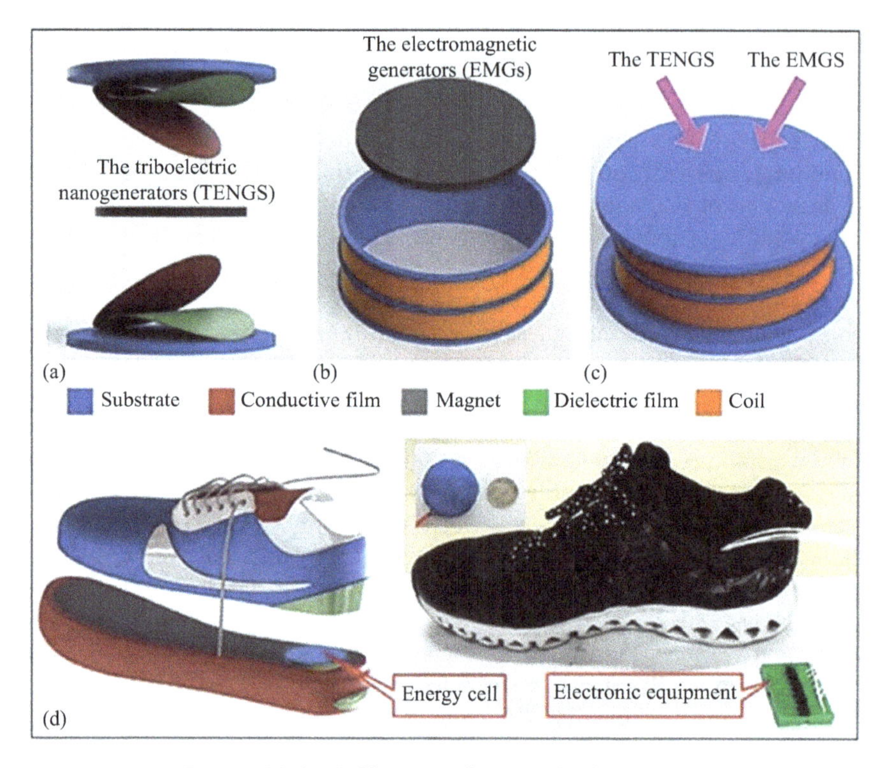


*Figure 10.6 Self-powered versatile shoes [309]*

metrics, including the power output, voltage, and current. Although the technology is still in the early stages of development, smart footwear has the potential to revolutionize the way we power wearable and mobile devices, making them more sustainable, efficient, and convenient. Biomechanical energy harvesting through smart footwear has numerous potential applications, especially in wearable medical devices. Further details about its applications are listed in Table 10.4.

## 10.7 Energy harvesting through a wristwatch

Energy harvesting involves capturing energy from a surrounding environment and converting it into usable electrical power. When it comes to human-centric energy harvesting, the challenge lies in the fact that the human body exhibits large amplitude movements at low frequencies, which makes it difficult to design an energy harvester that can efficiently capture and convert this energy [367]. To overcome this challenge, designers have proposed several approaches, and one such example is shown in Figure 10.7, with frequency up-conversion being one of the mainstream methods. This involves converting the low-frequency motion into higher-frequency signals that can be more efficiently harvested by transducer

*Table 10.4 Applications of smart footwear*

| Field/application | Description |
|---|---|
| Wearable medical devices | Energy-harvesting shoes could power sensors that monitor vital signs such as the heart rate, blood pressure, and respiratory rate, allowing for continuous monitoring of patients and early detection of health issues. |
| Sports and fitness | Energy-harvesting shoes could power activity trackers or GPS devices, eliminating the need for batteries and making the devices more lightweight and comfortable to wear. |
| Military | Energy-harvesting shoes could provide a sustainable and reliable source of power for soldiers' devices, thereby reducing the need for carrying multiple batteries. |
| Industrial | Energy-harvesting shoes could be used in industrial settings to power sensors and other small devices, thereby reducing the need for wiring and making the devices more flexible and adaptable. |
| Consumer electronics | Energy-harvesting shoes could be used to power small electronic devices, such as smartwatches, headphones, and other wearable devices, thereby reducing the need for frequent charging and increasing their sustainability. |

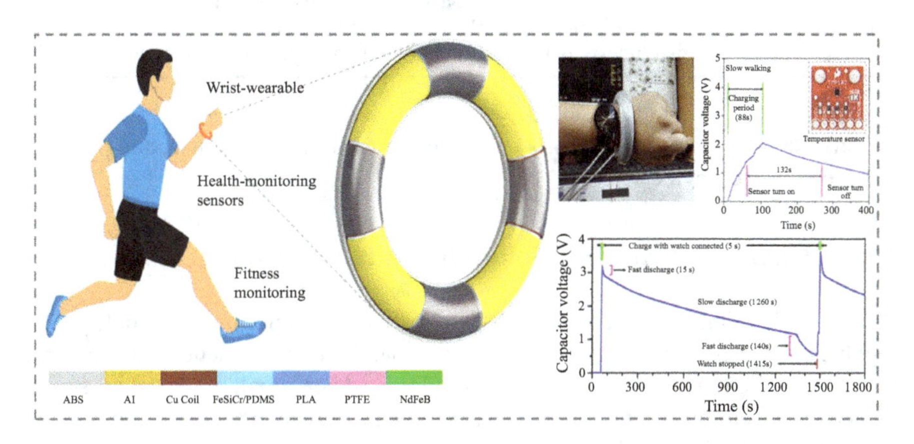


*Figure 10.7 Energy harvesting through a wearable watch [368]*

elements. In one example, researchers designed a miniaturized EMEH and integrated it into a wristwatch, which could scavenge energy from hand motion. The design included magnetic flux stacks, which increased the power density without increasing the magnetic volume of the device [224]. However, despite the progress made in designing smart wrist wearables that can harvest energy, the lack of a miniaturized device to integrate into wearable electronics remains a challenge.

There are several potential applications of biomechanical energy harvesting through wearable wristwatches. Table 10.5 shows different applications of biomechanical energy harvesting through wearable devices wristwatches. Applications

*Table 10.5 Applications of a smart wrist watch*

| Application | Description |
|---|---|
| Wearable medical devices | Biomechanical energy harvesting through wearable devices, such as wrist watches, can be used to power sensors that monitor vital signs such as the heart rate, blood pressure, and respiratory rate. This enables continuous monitoring of patients and early detection of health issues. |
| Sports and fitness | Biomechanical energy harvesting through wearable devices can be used to power activity trackers or GPS devices, eliminating the need for batteries, and making the devices more lightweight and comfortable to wear. |
| Military settings | Biomechanical energy harvesting through wearable devices can be used in military settings where soldiers need to carry a lot of equipment and rely on batteries to power their devices. |
| Smart homes | Biomechanical energy harvesting through wearable devices can be used to power smart home devices such as thermostats, security cameras, and smart locks. |
| Environmental monitoring | Biomechanical energy harvesting through wearable devices can be used to power environmental monitoring sensors that measure air quality, temperature, and humidity. |
| Industrial applications | Biomechanical energy harvesting through wearable devices can be used in industrial applications to power sensors that monitor machinery, detect malfunctions, and improve the overall efficiency. |

include medical monitoring, sports and fitness tracking, military use, and powering small electronic devices.

## 10.8 Energy harvesting through smart clothing

Biomechanical energy harvesting through smart clothing is a technology that involves capturing the energy produced by human body movements and converting it into electrical energy that can be used to power small devices or sensors embedded in clothing [369]. This technology can find several applications in different fields, such as military and outdoor activities, healthcare, and sports. In military and outdoor activities, smart clothing can power communication devices, GPS tracking systems, and environmental sensors, making them lightweight and more comfortable to wear. The technology can also be used in healthcare sectors to monitor the movement and posture of the elderly or disabled individuals for fall detection and prevention. The energy-harvesting technology in smart clothing works by incorporating flexible PE materials or EM generators into the fabric of the clothing (Figure 10.8). These materials can generate electrical energy from the mechanical stress or movement produced by the human body during activities such as walking, running, or even breathing [370–372].

Overall, biomechanical energy harvesting through smart clothing has the potential to revolutionize the field of wearable electronics by providing a

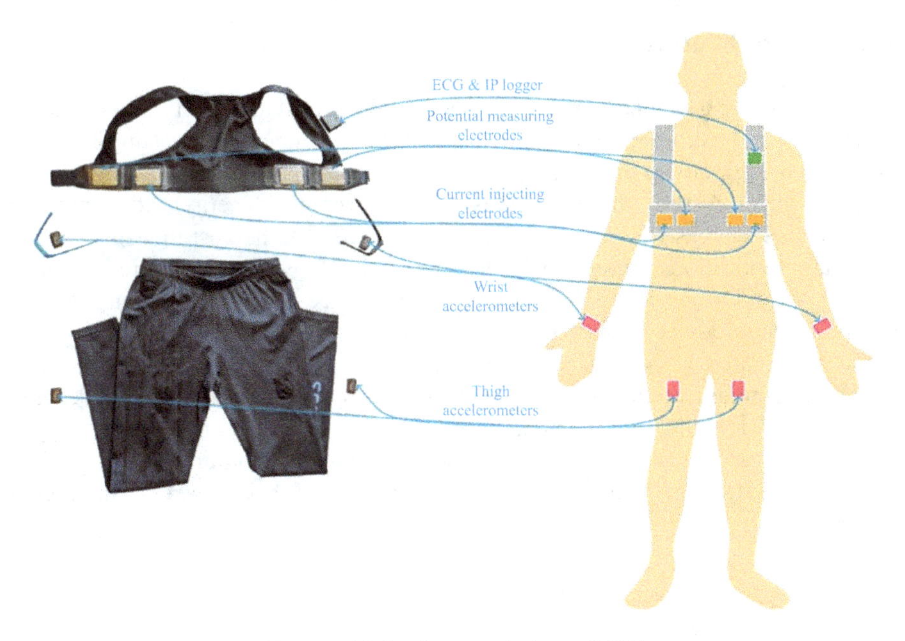


*Figure 10.8 Smart clothing [373]*

*Table 10.6 Applications of smart clothing*

| Application | Description |
|---|---|
| Military and outdoor activities | Powering communication devices, GPS tracking systems, and environmental sensors embedded in clothing for lightweight and comfortable use. |
| Healthcare | Monitoring movement and posture of the elderly or disabled individuals for fall detection and prevention. |
| Sports and fitness | Powering activity trackers, heart rate monitors, and other fitness devices to eliminate the need for batteries and increase comfort. |
| Fashion | Incorporating energy-harvesting technology into clothing to power wearable technology such as smartwatches or headphones. |
| Industrial and construction work | Powering sensors that can monitor the health and safety of workers, such as environmental sensors for air quality and temperature. |

sustainable and reliable source of power for small devices and sensors embedded in clothing.

Table 10.6 summarizes the potential applications of biomechanical energy harvesting through smart clothing, including military and outdoor activities, healthcare, and sports and fitness. Smart clothing technology can power sensors and devices embedded in clothing, making them lightweight and more comfortable to wear, while also improving the monitoring and tracking capabilities.

## 10.9 Conclusions

Energy-harvesting technology has the potential to provide a reliable source of power for small wearable devices, and biomechanical energy harvesting is a promising approach due to the abundance of kinetic energy from human motion. This chapter has discussed various sources of biomechanical energy and transducer technologies such as PE, EM, and TE generators used to harvest this energy. The potential applications of energy-harvesting technology were explored through various wearable devices like smart footwear, knee braces, wristwatches, and clothing. These devices can be used in healthcare, sports and fitness, military, and outdoor activities, providing continuous power for vital sign monitoring, activity tracking, communication, and environmental sensing. Although challenges remain, ongoing research and advancements in materials science and engineering suggest that energy-harvesting technology has the potential to revolutionize the field of wearable electronics, enabling a new generation of self-powered, intelligent devices.

*Chapter 11*

# Electromagnetic energy harvesters for space applications

Energy harvesting is a vital activity in almost all forms of space applications known to date, as it is not possible to bring, along with probe, spacecraft or rover enough supply of fuel to cover all the mission's energy needs. So far, many new concepts of mining or other form of manufacturing of fuel on exploration sites have been successfully performed; however, they are still based on an external energy source to generate fuel, and/or on initial energy to be harvested to trigger a self-sustaining cycle. This chapter summarizes key factors that determine satellite energy needs and explains the most widespread examples of energy-harvesting space devices, present in the Earth Orbit as well as explores the Solar System and beyond. Some of the discussed energy harvesters are suitable for very large probes in terms of the size and functionality, while other examples fit and scale easily on satellites ranging from dozens of tons to a few grams in weight.

## 11.1 Introduction

Energy harvesting is a concept of gathering energy from an ambient, often using a renewable energy source. Harvested energies are of various kinds, including being transmitted in EM waves of different lengths—from classic radiofrequencies used for communication to sunlight, thermal gradients, and kinetic and potential energies. Sources of energies that are being gathered are of two types. First is the intrinsic energy located in environment where an harvester operates (like the Sun providing power to photovoltaic (PV) cells). The second source is the other operating devices (like an electric motor providing power to a thermal or kinetic harvester), which dissipate the waste energy. Energy, when collected, is usually converted into electricity and then stored. Power generated by portable, autonomous harvester devices, depending on technology, used spans from ten of microwatts to tens of milliwatts. This fact implies that systems powered by energy-harvesting device must be small in terms of power needs. As a result, the functionality of such systems has also to be limited, especially when they are likely to operate in charge (energy build-up in span of minutes or tens of minutes) and act (perform the measurement and processing quickly and effectively).

State-of-the-art R&D boosted by increasing industrial interests in energy harvesting has already demonstrated as successful functional devices, including technological advances for automotive industry (tire pressure monitors using the PE effect), aerospace industry (fuselage sensor network in aircraft, using thermoelectric effect), manufacturing automation industry (various sensors networks, using thermoelectric, PE, electrostatic, and PV effects), personal devices (wristwatches, vital sign monitors, and remote controls, using PE and thermoelectric effects). However, these activities focus on miniaturized devices, consuming very low power and harvesting energy on a low scale, aiming to guarantee sensor autonomy, safety (no power lines), and minimal maintenance.

It is worth noting that those design drivers, which are listed above, are almost the same as baseline requirements for all human-made artificial Earth satellites and deep space probes exploring the Solar System and its boundaries, and recently, beyond (e.g., Voyager mission). Every satellite and/or probe is an energy-harvesting device, performing energy harvesting on a larger scale. Indeed, satellites have to operate continuously, and it is currently not feasible, from both technological and economical perspectives. The cost of 1 kg of mass sent to Low Earth Orbit (LEO) varies from 10k to 100k €.

Satellites and probes vary significantly in terms of the size, weight, and applications. Satellite energy needs, technology used for energy collection, and the source of harvested energy vary as well. On average, a satellite (or a probe, which is used interchangeably in this chapter) size can vary from several meters and several tons (large satellites, typically telecommunication, military earth observations, and telescopes) to a few centimeters and a few hundred grams (micro to nanosatellites, technology experiments, scientific, and amateur radio). Similarly, power needs to feed satellites varies from several kilowatts (for TV broadcasting relays) to hundreds of milliwatts (for satellite-on-chip experimental designs). Sources of energy for powering the probe are different, varying from the most popular PV solar cell arrays present in almost all satellites operating in the inner Solar System, to thermoelectric generators, working on temperature gradient between two heat storages, heated by solar infrared radiation (or radioisotopes) and cooled by cosmic background. Fuel cells and nuclear reactors were also employed as energy sources for satellites, but they are definitely not mainstream. Fuel cells were used in Space Shuttle, which had a limited operation time capability and are currently withdrawn from use, while nuclear reactors pose a serious threat to the environment in case of a launch failure, and due to heavy infrared signature, they are not used for military purposes. There is also another category of energy source used by spacecraft, and for micro- and smaller satellites, which are the electrodynamic tethers (EDTs).

Out of the energy sources in use in orbit are all based on harvesting energy forms which are widely present in LEO, including solar radiation (in visible and IR ranges), a gravity combined with Earth's magnetic and plasma environment. This chapter introduces the basic principles of operation of PV, thermoelectric, and tethered harvesters used or intended for use on board of probes, rovers, and satellites [374].

## 11.2 PV effect harvester

Earth-orbiting spacecrafts, ranging from LEO to geosynchronous orbit, typically employ solar PV as a source of energy. PV array implementation is efficient, simple, and scalable (up to 300 kW), and can be implemented on diverse classes of satellites. A PV harvester has a descent specific cost ($/W), offers unlimited access to "fuel," and has minimal safety analysis reporting needs.

Solar arrays are assembled from a large number of individual solar cells arranged on mechanical supporting panels, which convert solar energy into electric power, by PV conversion. The solar cells are designed in various shapes and sizes, and put out together a relatively low individual current and voltage. The first space application of solar PV power dates back to March 17, 1958, when Vanguard 1 was launched, utilizing six solar panels that successfully provided less than 1 W of power for over six years, with approximately 10% of power conversion efficiency. Solar cell design is rated by its ability to convert a certain percentage of solar energy into electric power, which is known as the solar cell efficiency, which is defined by (11.1):

$$n = P_{\text{out}}/P_{\text{in}} \tag{11.1}$$

where $P_{\text{out}}$ is the electrical power output and $P_{\text{in}}$ is the solar power input.

Solar cells are organized to cover as much area as possible, to exhibit as much of insolation as possible. The solar constant, which is equal to 1 358 W/m$^2$, is the total solar energy incident on a unit area perpendicular to the Sun's rays at the mean Earth–Sun distance outside the Earth's atmosphere. It varies between about 1 310 and 1 400 W/m$^2$, depending on the annual cycle, with a maximum at perihelion and a minimum at aphelion. Solar cells can take most of this power when they are set at 0° with respect to the incoming solar rays. Increasing this angle will lower the effective irradiation power trigonometrically, and this is precisely the reason for implementing a sophisticate attitude determination and control systems on board of larger spacecraft, to ensure that solar arrays take most out of incoming light with respect to the current orbital operations during the satellite mission.

Solar cells are connected in series to maximize the voltage and in parallel for current maximization. To minimize the power losses with a single cell failure, the solar array cells are connected in a series parallel ladder network.

It is critical that solar arrays collect enough energy during the sunlight period to power the spacecraft during the entire orbit (eclipse time can take around 40%–45% of the whole orbital period).

The cell design is affected by various factors which must be considered, such as *I–V* characteristics, its temperature dependence, distance from the Sun, and the radiation degradation. The current–voltage (*I–V*) characteristics of solar cells are hence of crucial importance in the design of solar arrays. PV arrays are aimed to be designed for minimum mass and maximum efficiency at the maximum power point (MPP). The MPP is where the product of $I$ and $V$ is at a maximum, which is defined by the maximum area rectangle within the plot. This point always falls at the knee of the *I–V* curve (Figure 11.1) [374].

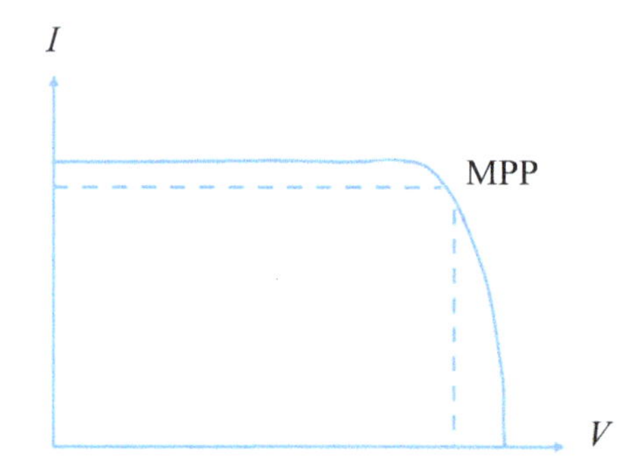


*Figure 11.1 PV cell current–voltage curve*

There are two methods generally employed for working with the MPP of a PV array. The first method is typically used for smaller satellites and involves using direct energy transfer (DET) control. The DET systems dissipate unnecessary power and typically use shunt resistors to maintain the bus voltage at a predetermined level. Shunt resistors are usually at the array or external banks of resistors and employed to avoid internal heating. It means that solar array has to be sized in such a way to cover worst-case insolation case and worst-case satellite power needs (and to recharge batteries for eclipse).

The second method is more sophisticated and dynamic, and consists to implement a peak power tracker (PPT) distribution system. PPTs extract the exact power required from the solar arrays. Some DC/DC converters have to be designed in series with the array in order to dynamically change the solar array's operating point—according to changing the system needs. Typically, it requires 4%–7% of the solar array power to operate and is often applied on larger satellites [375].

## 11.3 Thermal energy harvesters

In space, there are two types of harvesters transforming heat into electricity, which are largely used. The first type is based on a static thermal energy harvester, typically a thermoelectric generator built around a radioisotope source. The second type of power generator subsystem, often applied in GPS constellation satellites, is based on an alkaline electrochemical cell, which is an example of dynamic thermal energy harvester.

The physical phenomenon that serves as a base for the creation of a thermoelectric harvester is the Seebeck effect. Seebeck effect is a process of direct heat gradient into electricity conversion over a two different materials junction. If it is assumed that one material is in a given temperature and the other one forming the junction is in another level of temperature, then the resulting junction voltage is proportional to the difference between the two (i.e., the gradient). The proportion is a characteristic of the material employed and is called the Seebeck coefficient (thermopower), which is defined as:

$$\alpha = \Delta V / \Delta T \quad (11.2)$$

The Seebeck coefficient value may vary significantly and ranges from ~1 mV/K (for metals) to ~100 mV/K (for semiconductors). The Seebeck effect has also a reversed counterpart called Peltier effect, also widely used in space applications—for refrigeration purposes, as cooling of some electronic part is heavily limited due to vacuum and lack of convection, which constitutes for 80% of heat transfer when it takes place under terrestrial conditions. Similarly, Seebeck effect is used in all thermoelectric generators that are customized for operation in outer space.

Figure 11.2 shows a simple model of a Peltier cell (left panel) and a Seebeck cell (right panel) built on semiconductor chunks. When heat is absorbed on the top side of a thermoelectric generator, the movable charge carriers begin to diffuse, creating a uniform concentration distribution in the device along with the temperature gradient, and producing the difference in the electrical potential on both sides of it. As it has been explained, due to the Seebeck effect, electrons flow through the n-type semiconductor chunk to the colder side, while in the p-type semiconductor, the holes flow to the colder side. To maximize the power generation output, thermoelectric cells are connected in chains (thermally in parallel and electrically in series) to harvest as much heat as possible [376,377].

Most of the space applications of thermoelectric generators assume coupling with some radioisotope heat source. In fact, the NASA has designed such a standard unit called radioisotope thermoelectric generators (RTGs). During the time of writing this chapter, 41 RTGs have been launched on 23 spacecrafts. Their power production ranges from 2 to 300 W. The total system efficiency is around 7% (~ 2 kW heat production in order to obtain 130 W of usable spacecraft power). The RTGs use plutonium-238 with an 87.7-year half-life [374].

To provide higher efficiencies of electric power production, the development of space solar dynamic power systems has been proposed. The difference between

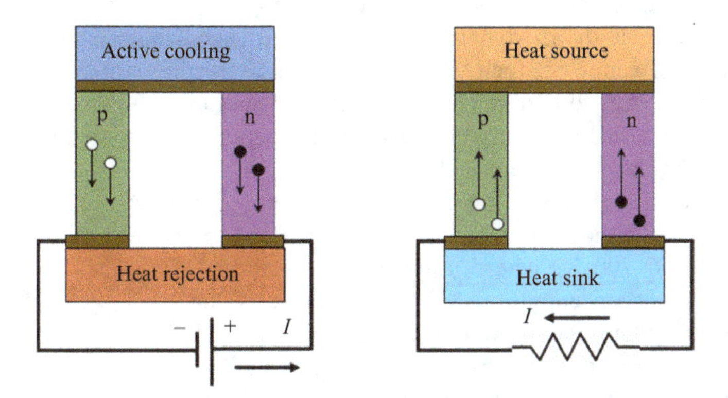


*Figure 11.2 Semiconductor thermocouples depicting the Seebeck and Peltier effects*

thermal static and thermal dynamic power is the power conversion technique. Instead of the direct conversion of heat into electricity as with thermoelectric effect, solar dynamic systems use solar power to heat a working fluid to drive a heat engine, which is used to generate electricity. The advantage of solar dynamic systems over solar PV systems is that dynamic systems in general have a higher thermal efficiency and can be used for higher power levels. A solar dynamic system consists of four basic components, the collector/concentrator, receiver, radiator, and thermal storage material, and the heat engine turning heat into movement. Movement is then turned into electricity in an alternator. The power conversion cycle can be any of the common thermodynamic cycles: Rankine, Brayton, or Stirling. A similar thermal dynamic power generator, but working on a slightly different principle, is an electrochemical concentration cell that converts the work generated by thermal expansion of sodium vapor directly into electric power, called AMTEC (alkali metal thermal-to-electric conversion).

Solar AMTEC (SAMTEC) is a new power system concept implemented recently in new generation of GPS satellites. The SAMTEC system consists of array of multitube, vapor anode AMTEC cells (having 24% conversion efficiency), and a direct solar irradiation receiver integrated with LiF salt canisters performing a role of energy storage even during maximum eclipse periods. Due to effective heat energy storages, electrical battery packs storage capacity is much less necessary. AMTEC cells are connected in series and in parallel, and each cell is redundant in order to achieve seamless 10–12 years in orbit operation without the reduction of energy conversion efficiency owing to the background radiation, which could be a serious issue in case of PV arrays. AMTEC cells are the heart of the described harvester. The device accepts a heat input in a range from about 900 to 1 300 K, and produces direct current with an expected efficiency between 15% and 40%. In the AMTEC cell, sodium is flowing around a closed thermodynamic cycle between a high-temperature heat reservoir (salt canister) and a cooler reservoir at the heat rejection temperature (radiator fins). The unique characteristic about AMTEC cycle is that sodium ion conduction between a high-pressure or an activity region and a low-pressure or an activity region on either sides of a ion-conducting refractory solid electrolyte, is thermodynamically nearly equivalent to an isothermal expansion of sodium vapor between the same high and low pressure. Electrochemical oxidation of neutral sodium at the anode leads to sodium ions, which traverse the solid electrolyte, and electrons that travel from the anode through an external circuit where they perform electrical work, to the low-pressure cathode, where they recombine with the ions to produce low-pressure sodium gas. The sodium gas generated at the cathode then travels to a condenser at a heat rejection temperature of perhaps 400–700 K, where liquid sodium reforms. The AMTEC thus is an electrochemical concentration cell which converts the work generated by the expansion of sodium vapor directly into electric power [378].

A typical SAMTEC electrical power supply configuration consists of a symmetric, rigid parabolic concentrator supported by rods extruding from the front face of a solar receiver. Each receiver consists of AMTEC cells and TES canisters, short sodium heat pipe, and a heat receiver plate (called a hot shoe). Cells and canister assemblies are arranged in such a way so as to form the solar receiver cavity walls. Solar radiation enters through aperture and falls mainly on the sidewalls, and heats the canister and sidewalls of the AMTEC cells. Heat receiver plates are designed in such a way to absorb incident solar energy and then radiate part of it to lower parts of the solar receiver cavity. Thermal gradient is formed along the main axis. Such a device provides around 1 200 W of power available to spacecraft bus [379,380].

## 11.4 Electrodynamic tether harvester

A space tether is a long thin structure (line like) that extends from a suborbital or orbiting spacecraft such as a rocket, satellite, or a space station. Space tether length can vary from single meters up to 30 or so km (the world record is more than 32 km). Tethers must be of extraordinary strength to withstand tensile forces. For tethers, three categories of applications could be distinct: momentum exchange, formation flying, and electrodynamic.

Mechanical tethers (first two categories) connect masses on the orbit but do not have to be conductive. EDTs are conductive. An EDT can be built using any conductive material, and the tradeoff between the mass and resistivity is a matter of case-by-case analysis (typically, it is aluminum or copper).

EDT operation principles are straightforward. First, as the tether moves along its orbital path, there is an EMF generated along it. Second, tether wire is a low-resistance path connecting regions of different plasma density and parameters. Third, connection to ionosphere can be limited only to tether ends or can be continuous along the tether length, which affects the electrodynamic properties of the whole system.

EMF is a Lorentz force that results from tether (long conductor) movement in the Earth's EM field (11.3).

$$F = q(E + v \times B) \tag{11.3}$$

where $q$ is the charge of an electron, $E$ represents any ambient electric field (small), and $v \times B$ represents the motional electric field as the tether travels at a velocity through the Earth's magnetic field, represented by $B$. Equation (11.3) can be rewritten as:

$$F = qE_{\text{tot}} \tag{11.4}$$

where

$$E_{\text{tot}} = E + v \times B \tag{11.5}$$

and represents the total electric field along the tether.

To get the total EMF generated along the tether, $E_{\text{tot}}$ must be integrated along the entire length of the tether. The resulting total tether (of length $l$) potential is:

$$\theta_{\text{thether}} = -\int_0^l EE_{\text{tot}} dl \approx -\int_0^l v(l) \times B(l) dl \tag{11.6}$$

which is negative as electrons in the tether are acted upon by the Lorentz force. As the ionosphere plasma surrounding the EDT decent conductor and ambient electrostatic field is small, the tether potential can be ignored.

The tether potential is path independent assuming a conservative resultant electric field and steady-state conditions. As a result, the value of $\theta_{\text{tether}}$ depends only on the relative locations of the wires endpoints (their separation distance and orientation) and does not depend on the position of the tether between both ends.

The second and third principles are related to current flow through the tether, which occurs when a connection is made between the wire ends and orbital plasma. This can be done passively or enforced actively. A passive solution requires voltages in the tether–plasma system distribute themselves in a self-consistent manner, which can require a very high level of tether charging in order to enable current flow [381,382].

An active solution requires an electron generator device, such as an electron emitter or a hollow cathode. Both solutions cause current flows through the wires as shown in the schematic of Figure 11.3. In the passive solution, current flows up the tether because the resultant force on the electrons is downwards (towards the Earth). After electrons are collected at the remote end of tether, where the counter mass, a subsatellite is located, they are conducted through the tether to main spacecraft where they are ejected into the surrounding plasma. Current closure occurs in the ionosphere, completing the electrical circuit.

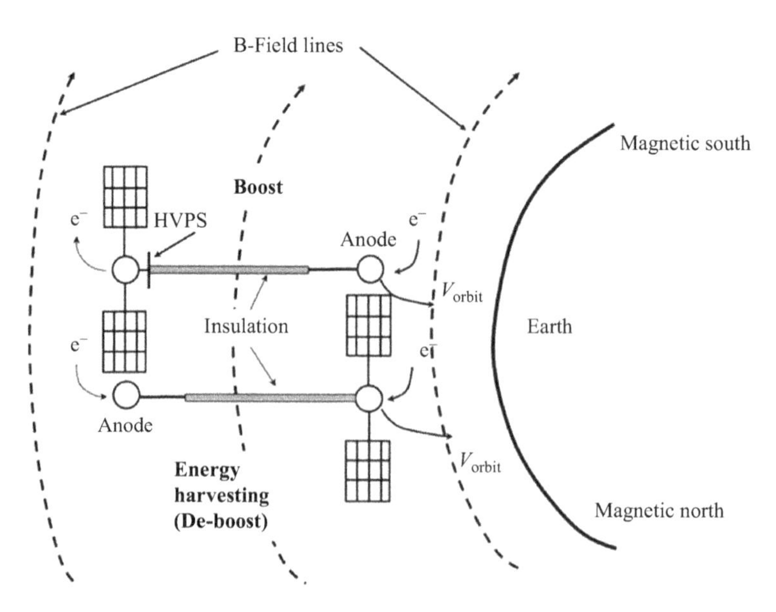


*Figure 11.3 EDT in energy-harvesting and -boosting modes*

If current flows in the tether element, a force is generated and can be expressed as:

$$F = \int_0^l I(l) \times B(l) dl \tag{11.7}$$

When tether is in energy-harvesting (or, in other words, de-orbit mode), the EMF can be used by the tether system to supply the current from the tether into electrical loads, including energy-storing devices like flywheels and batteries, eject electrons at the emitting end or grab electrons at the counterbalance end. In the boost mode (orbit-raising mode), on-board power supplies must overcome the EMF resulting from travel through magnetosphere, to drive current in the opposite direction, thereby creating a force in the opposite direction than the Earth, boosting up the system. Thrust levels vary significantly on applied power and typically are less than 1 N, which is still three to six orders of magnitude better than those generated by other electric propulsion like ion or Hall-effect engines.

Numeric simulations, verified by experimental results, show that for a large satellite, a 10 km-long EDT can provide an average power of 1 kW (2 kW in peak) with 70%–80% efficiency of potential to electric energy conversion. Typically, such a tether, if made out of aluminum, weighs almost 8 kg, which is almost negligible when compared to thousands or more kilograms for a large Earth-orbiting satellite [383,384].

Smaller satellite platforms are also considered as hosts for EDT, namely nanosatellites called Cubesats. Cubesats, depending on configuration, typically does not exceed 4 kg and 30 cm in the largest dimension for a 3-unit model. An experimental prototype has been built, allocation 1.33 $kg^{-1}$ unit for all EDT components, including end-mass, spool, deployment controller, and an electron gun. For the tether, a low-mass variant of AWG 25 aluminum wire has been selected. A length of 1 300 m (resulting in tether mass bit below 0.6 kg) is a breakeven point, where the tethered system can generate the same amount of power as state-of-the-art PV array of size that could be reasonably accommodated on a 3-unit Cubesat.

The extraordinary growth of nanosatellite applications has facilitated increase in interest of even smaller spacecraft, pico-satellites, satellites-on-chip operating in distributed swarm. Advances in electronics design and MEMS paving their way on board of pico-satellites have proven the feasibility of construction and applicative usability. Problem, although, lies in orbital dynamics. Satellites-on-chip suffer from low ballistic coefficients. They have very low mass (thus, kinetic energy) while having a relatively high area of average cross-section (thus, aerodynamic drag). As a result, their orbital lifetime ranges from days to hours. As a solution, to counter the drag force, a short, semirigid EDT for propulsion is considered, which keeps the mass of the overall satellite-on-chip rather low and efficiently provides enough thrust to overcome drag in LEO. Recently, the Defense Advanced Research Projects Agency (DARPA) performed an experimental flight of two pico-satellites (100 g, 5 cm $\times$ 5 cm $\times$ 2 cm) connected with 30-m long EDT. In course of mission, a satellite–tether–satellite system performed scheduled operations, including the boosting orbit altitude, energy harvesting, and some more sophisticated formation flight maneuvers. All maneuvers were satisfactory to an extent which led the scientist to consider tethered femto-satellite swarms (1 g, 1 cm × 1 cm × 1 cm) [385–387].

## 11.5 RF energy harvester optimized for WSN in space launcher applications

Historically, developments on wireless energy harvesting have demonstrated the ability to power many systems. The aim is to harvest the energy transmitted by a high-frequency transmitter and then to adopt it to power a system. This technology can be used in smart electronic devices, environmental sensors, and in our case, spatial telemetry.

The energy recovery unit is integrated into a network of RFID sensors [388], capable of rapidly measuring and transmitting the temperature in a launcher equipment box. Each sensor offers a high level of flexibility in terms of periodicity measurement and data transmission. It is possible to transmit the data in packets of seven units, with a periodicity of measurement as low as 10 ms. This variation of operation implies a consumption of between 5 mJ and 50 μJ, depending on the periodicity of measurement. It also offers the possibility of keeping a time stamp of the measurements as long as there is sufficient energy to carry out a complete measurement and transmission cycle.

The system highlighted in this section was specially developed to integrate into a launcher equipment box. The diameter of a box is about 5 m and is completely metallic. The system is adapted to this metal environment and allows operation on at least one-half of the box. The maximum distance, measured as a semicircle, was about 8 m with metal obstacles. At such distances, the received field strength is very low and entails an initial start-up time as well as restrictions on the utilization of the recovered energy.

The needs of this domain combined with the environment system have led to changes with respect to a typical architecture [389]. A second reception chain was added (Figure 11.4) in order to increase the energy received and the storage capacities. An energy-management function has also been developed to allow the distribution and maximization of system power times.

The following section presents an RF harvesting solution enable to supply an UHF RFID-based tag sensor developed by Collignon *et al.* [390]. The solution is integrated in a WSN specifically designed for launcher applications. To achieve a complete autonomous sensor network and drastically reduce the risk of explosion or contamination caused by chemical battery, the harvesting-powered solutions are fully appropriate. With the low power required by the sensor tag (<50 μJ), the RF harvesting design allows supply over 5 m with an adaptive power management over the RF energy harvested. The purpose design includes a dual-input module to optimize the efficiency over the range of operation.

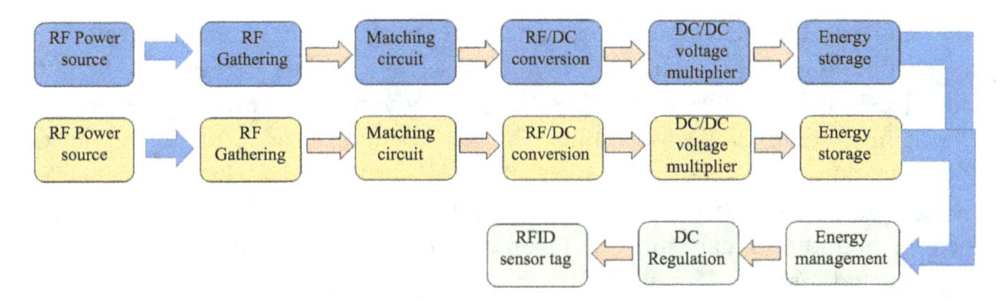


*Figure 11.4 Purpose RF energy-harvesting architecture*

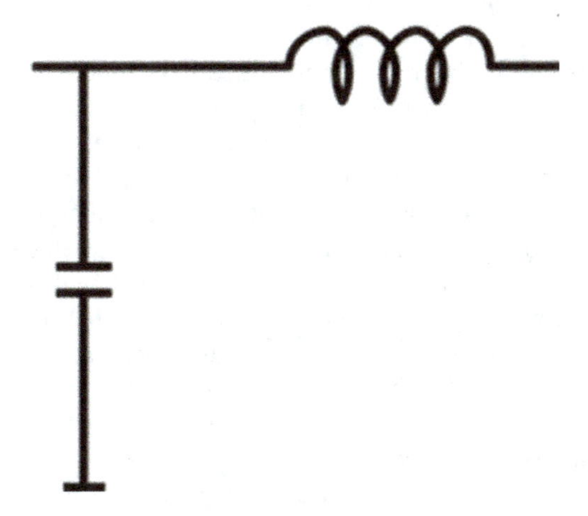

*Figure 11.5 Matching network*

## *11.5.1 Design description*

Two dipole antennas engraved on a PCB have been used. The advantage of using a dipole is that one can obtain a sufficiently wide bandwidth so that the performances are not impacted by a mismatch due to the metallic environment. The maximum gain obtained is of the order of 0 dB for a bandwidth of 68 MHz. The matching circuit maximizes the transmitted power from the antenna to the RF/DC converter. Thus, it is necessary to adapt the impedance of the antenna to the conjugate of the RF/DC converter for the frequency of interest. Since the frequency of RFID is lower than 1 GHz, a simple low-pass matching with discrete components was adopted (Figure 11.5). However, the nonlinearity of the RF/DC converter and the variation of the load on the DC/DC converter make it difficult to adapt perfectly over the entire range.

An average impedance measurement was performed for a frequency of 868 MHz. These impedance measurements made it possible to identify a network maximizing the output voltage of the RF/DC converter and a good matching to a 50 Ω antenna. The RF/DC conversion must make it possible to rectify the RF voltage and to multiply the DC voltage level. There are many assembly options available to perform this function, for example, CMOS-based assemblies [391,392] or Schottky diodes [393–395]. CMOS assemblies have the advantage of offering better sensitivity compared to diode assemblies. In this specific case, a diode circuit was adopted made with HSMS2852, which offers greater flexibility. Indeed, the number of stages of the assembly has an impact on the sensitivity and performance of the function. In addition, care must be taken to vary the performance as a function of the connected load on the output.

## *11.5.2 Design performance*

The use of two reception chains has several advantages. Due to the use of two antennas, the power at the input of the harvesting circuit is increased, and the probability of a fading problem due to multipaths is reduced. A yield comparison between the DC output and the received radiated power (Figure 11.6(a)) shows that the use of the two inputs leads to a gain in performance. In addition, the output power of the system from a received power of 10 dBm (Figure 11.6(b)) is greatly improved. Finally, the startup times, inherent of the first charging of the capacitors and the storage capacity of the harvester, were improved when compared to the use of a single channel. The measurement was performed with conductive sources connected on the harvesting inputs with the assumption that the received power is the same on the two inputs. The output measurement was achieved on the load point when a 3 V regulated voltage is near to drop. The maximum performance achieved with a single channel is 25% for a received power of 0 dBm. For this measurement, authors have used the most sensitive receiving chain, i.e., the one that has a DC/DC voltage booster. When using the two measurement chains, a performance of more than 25% over a received power range of −2 to 11 dBm is achieved, due to the use of a different sensitivity between the two reception chains.

The output power (Figure 11.6(b)) is taken with an output voltage set at 3 V, which corresponds to the supply voltage of the measuring sensor. Up to 5 dBm of received power, the output power does not exceed 1 mW. When the received field is higher, the second reception chain offers better performance and reaches 6 mW for 15 dBm. The use of the two inputs with different levels of performance makes possible to optimize the sensitivity at low power and the efficiency for the highest

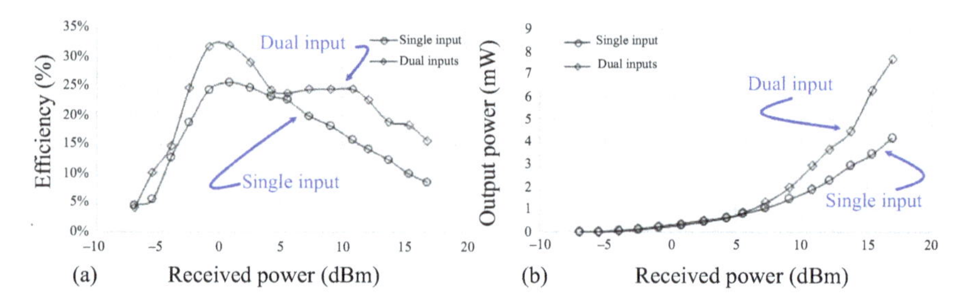


*Figure 11.6 (a) Power efficiency and (b) output power*

*Figure 11.7 Sensor with an incorporated harvester*

powers. This allows system operation over the entire power range under consideration with optimized power supply intervals of the sensor. One single temperature measurement requires only 50 μJ, and its interval of measurement is configurable from one to 100 per second, so that the maximum consumption was 5 mW. When carrying out the tests in the equipment box (Figure 11.6(b)), two modes of operation of the system could be tested. A permanent mode of operation, which allows the measurement sensor to keep a timestamp as well as continuity of measurements, and an intermittent mode of operation, which allows operation at lower powers, at the cost of power cuts.

During the tests, transmit measurements up to 8 m of semicircle length in the intermittent mode were possible, which correspond to half of the equipment box. To cover the entire test environment, only two RFID transmitters should be installed. The permanent mode works up to a little less than 5 m, limiting the periodicity of measurement to 5 s. However, in some areas, the received field does not allow the operation of the harvester and this phenomenon occurred mainly when a metal object obstructs the measurement system (Figure 11.7).

## 11.6 Conclusions

An energy harvester had its very important place in space technologies long before they started to pave their way into consumer, automotive, industrial control, and biomedical applications. Even if well described from a theoretical point of view and well simulated, and with decades of heritage of successful exposure to the most severe environmental conditions, space energy harvester designs do not develop quickly and do not evolve in different directions as their terrestrial counterparts. It seems that, once again, part of industrial and political activity which should be the avant-garde of technology, science and curiosity—space exploration is at the same time one of most conservative of all human activities. Space-related conservatism is originating at reluctance to take risk, especially when it paired with high cost—of components, of qualified work force and of launch services. At least dawn of nanosatellites (and smaller ones), bring hope that some of the innovative solution, especially tether based, and will pave their way into day-to-day satellite operations, enabling more power generation and therefore more diverse and sophisticated spacecraft functions, and longer orbital lifetimes.

*Chapter 12*

# Conclusions and outlook into the future

## 12.1 Conclusions

Energy harvesting for remote sensing and flexible electronics through hybrid technologies is a rapidly growing field that has the potential to revolutionize the way we collect and transmit data in various industries. From healthcare to aerospace, energy-harvesting technologies offer a sustainable and cost-effective solution to powering sensors and electronic devices. In this book, we have explored various hybrid technologies for energy harvesting, and we have also discussed the challenges and opportunities associated with implementing these technologies in real-world applications. Overall, we have seen that energy-harvesting technologies have come a long way in recent years, with significant improvements in efficiency, reliability, and scalability. However, there is still much room for improvement, particularly in terms of the increasing power density, reducing cost, and optimizing energy storage.

More specifically, low-power wireless microsensors and microelectronic devices have become a very popular research topic in recent years. This is because the ambient energy obtained from vibrations of moving bodies, like humans, machines, vehicles, bridges, buildings, ocean waves, and other civil structures, provides a continuous and efficient alternative to low-power devices, thus providing a replacement to batteries. The batteries have a limited lifespan and they must be replaced repeatedly, which precludes the use of WSNs in embedded and remote locations. Biomechanical energy harvesting can be an everlasting alternative for the sustainable operation of microsensors and portable microelectronic gadgets for the nonstop health-monitoring applications of well-being parameters. Harvesting energy from human motion can boost up the performance and reliability of wearable devices by making them highly self-sustainable. In this thesis, a compact, lightweight nonlinear EM, hybrid PEM-IEH has been developed to successfully harvest low-frequency walking energy into useful electrical energy for the use in low-power microelectronics and embedded systems.

In light of the technical background of IEH parameters, the length of the cantilever beam, resonant frequencies, dimensions, and weight of the harvesters were considered as important parameters for the design of insole energy harvesters, due to the constraint volume of the insole, and these concepts are discussed in Chapters 1 and 2. In Chapter 3, a relationship between the beam length and resonant frequencies within given constraints was established. This was later used to design and fabricate EM and HEHs in the later chapters.

In Chapter 4, the designed harvester was modeled as a mass–spring–damper system and simulated for Eigenfrequencies analysis in COMSOL Multiphysics® for predicting the resonant modes of the EMIEH. The spring–magnet assembly was developed as a 3D model in PTC Creo™. Based on the 3D model, the circular spiral spring was fabricated from commercially available galvanized steel using a CNC-EDM wire-cut machine. The 12 mm disk magnets were pasted on the top and bottom sides of the central platform of the spring as tip-proof masses to reduce its resonant frequency. Teflon spacers were used to have a gap between the magnets and coils during free oscillations of the spiral spring during external vibrations. The harvester was characterized for electrical output inside the laboratory and incorporated into the sole of a shoe for harvesting biotechnical energy harvesting during the foot strides. An average total power of 839 $\mu$W was produced across both the upper and lower wound coils at a resonant frequency of 8.9 Hz under 0.6 g base acceleration. The EMIEH was installed inside the shoe, and a 100 $\mu$F capacitor was charged up to 1 V from normal foot movement for about 8 min during walking.

The designed EMIEH in Chapter 4 was further optimized for improved power generation to a hybrid device generating power by combined PE and EM transduction simultaneously in Chapter 5. The hybrid PEM-IEH was fabricated by using piezoceramic plates for piezoelectricity, and the number of coils turns in the wound coils was increased with a greater number of magnetic field lines and effective EM induction. The hybrid PEM-IEH generated a total power of 1 400 $\mu$W across the upper and lower EM generators and a total power of 269 $\mu$W across the piezoelectric units under 0.5 g base acceleration at resonance. The harvester charged a 100 $\mu$F capacitor up to 2.4 V by combined PE-EM transduction across the upper and lower hybrid generators.

In Chapter 6, a HIEH based on PEM-IEH was presented. The developed prototype is a square spiral spring-type multi-degrees-of-freedom with multiresonant frequencies and improved output performance. The harvester exhibits six resonant frequencies and produced about 7 V peak–peak voltage from the upper and lower hybrid generators, which is more than that of the EMIEH and PEM-IEH, as discussed in Chapters 4 and 5, respectively. Flexible PVDF cantilever beams were used as PE generators which adds extraresonant frequencies to the harvester and hence more voltage and power peaks. The HIEH charged a 100 $\mu$F capacitor up to 3 V by both the piezoelectricity and electromagnetism from walking strides for about 8 min. The energy harvesters developed in this research work are compact sized, lightweight, low fabrication cost, multiresonant, and can easily be incorporated into the sole of a commercial shoe for harvesting walking energy into useful electrical power. These devices have shown sufficient power-producing capability and can be easily integrated with the microsensors and microelectronic gadgets for real-time human physical monitoring applications. In Chapter 7, focus was put on the design, modeling, fabrication, and characterization of hybrid PEM-IEHs, where the architecture, working mechanism, and finite element modeling were detailed along with the fabrication of prototypes and the needed experimental setup and its associated experimental results. Chapter 8 is dedicated to the multi-degree-of-freedom hybrid PEM-IEHs, and includes the design and finite element modeling, the structural and electromechanical models, the fabrication setup, and the corresponding experimental results. In Chapter 9, we overviewed the FEA and its applications in KEH devices, while Chapter 10 highlights the energy harvesters for biomechanical applications. Finally, Chapter 11 concludes the book with "Electromagnetic energy harvesters for space applications," where two main applications were detailed, namely the electrodynamic tether harvester and the RF energy harvester optimized for a WSN in space launcher applications.

## 12.2 Future recommendations

Looking ahead, we can expect to see continued advancements in energy-harvesting technologies, driven by the ongoing research and development efforts. Here are some of the key trends and opportunities that we can expect to see in the future of energy harvesting.

Increased efficiency and power density: One of the biggest challenges facing energy-harvesting technologies is improving their efficiency and power density. Researchers are working on developing new materials and designs that can improve the conversion efficiency of energy-harvesting devices, as well as increasing their power output. In addition, advancements in nanotechnology include opening up new possibilities for energy harvesting, including the development of ultrathin and flexible materials that can harvest energy from a variety of sources.

Integration with the IoT: The IoT is rapidly transforming many industries, and energy-harvesting technologies are playing a key role in enabling this transformation. With proliferation of connected devices, there is a growing need for sustainable and scalable power solutions, where energy-harvesting technologies can fulfill the needs. In the future, we can expect to see an increased integration between energy-harvesting devices and IoT platforms, thereby enabling more efficient and intelligent energy management.

Applications in healthcare: Energy-harvesting technologies have the potential to revolutionize the healthcare industry, enabling more personalized and continuous monitoring of patient health. For example, wearable devices that harvest energy from the human body can provide a noninvasive way to monitor vital signs, such as the heart rate and blood pressure, without the need for batteries. In addition, energy-harvesting technologies can be used to power implantable devices, such as pacemakers and sensors, reducing the need for frequent surgical replacements.

Applications in aerospace: Energy-harvesting technologies are also finding applications in the aerospace industry, where they can be used to power remote sensors and other electronic devices on spacecraft and satellites. This can enable more efficient and cost-effective data collection, as well as reducing the need for frequent battery replacements. In addition, energy-harvesting technologies can be used to power unmanned aerial vehicles (UAVs) and other autonomous systems, enabling longer flight times and greater mission capabilities.

Advancements in energy storage: Finally, advancements in energy storage technologies are critical to the future of energy harvesting, enabling more efficient and reliable energy management. Researchers are working on developing new materials and designs for batteries and supercapacitors, as well as exploring new energy storage options such as hydrogen fuel cells and flow batteries. These advancements will be critical to unlocking the full potential of energy-harvesting technologies and enabling their widespread adoption.

Overall, the future of energy-harvesting looks bright, with significant opportunities for innovation and growth. As energy-harvesting technologies continue to advance, we can expect to see them play an increasingly important role in powering the devices and systems of the future. Irrespective of healthcare, aerospace, or other industries, energy harvesting has the potential to transform the way we collect and transmit data, enabling more sustainable and cost-effective solutions for a wide range of applications.

# References


[1] Siang J, Lim MH, Salman Leong M. Review of vibration-based energy harvesting technology: Mechanism and architectural approach. *Int J Energy Res* 2018;42:1866–93. https://doi.org/10.1002/er.3986.

[2] Fan K, Cai M, Liu H, Zhang Y. Capturing energy from ultra-low frequency vibrations and human motion through a monostable electromagnetic energy harvester. *Energy* 2019;169:356–68. https://doi.org/10.1016/j.energy.2018.12.053.

[3] Khan FU, Iqbal M. Development of a testing rig for vibration and wind based energy harvesters. *J Eng Appl Sci* 2016;35:101–10.

[4] Litak G, Friswell MI, Adhikari S. Magnetopiezoelastic energy harvesting driven by random excitations. *Appl Phys Lett* 2010;96. https://doi.org/10.1063/1.3436553.

[5] Zhao C, Zhang Q, Zhang W, *et al.* Hybrid piezo/triboelectric nanogenerator for highly efficient and stable rotation energy harvesting. *Nano Energy* 2019;57:440–9. https://doi.org/10.1016/j.nanoen.2018.12.062.

[6] Zhang Y, Zheng R, Shimono K, Kaizuka T, Nakano K. Effectiveness testing of a piezoelectric energy harvester for an automobile wheel using stochastic resonance. *Sensors (Switzerland)* 2016;16. https://doi.org/10.3390/s16101727.

[7] Rostami AB, Armandei M. Renewable energy harvesting by vortex-induced motions: Review and benchmarking of technologies. *Renew Sustain Energy Rev* 2017;70:193–214. https://doi.org/10.1016/j.rser.2016.11.202.

[8] Zhang B, Song B, Mao Z, Tian W, Li B. Numerical investigation on VIV energy harvesting of bluff bodies with different cross sections in tandem arrangement. *Energy* 2017;133:723–36. https://doi.org/10.1016/j.energy.2017.05.051.

[9] Usman M, Hanif A, Kim IH, Jung HJ. Experimental validation of a novel piezoelectric energy harvesting system employing wake galloping phenomenon for a broad wind spectrum. *Energy* 2018;153:882–9. https://doi.org/10.1016/j.energy.2018.04.109.

[10] Karami MA, Inman DJ. Powering pacemakers from heartbeat vibrations using linear and nonlinear energy harvesters. *Appl Phys Lett* 2012;100. https://doi.org/10.1063/1.3679102.

[11] Wei C, Jing X. A comprehensive review on vibration energy harvesting: Modelling and realization. *Renew Sustain Energy Rev* 2017;74:1–18. https://doi.org/10.1016/j.rser.2017.01.073.

[12] Izhar, Iqbal M, Khan F. Hybrid acoustic, vibration, and wind energy harvester using piezoelectric transduction for self-powered wireless sensor node applications. *Energy Convers Manag* 2023;277:116635. https://doi.org/10.1016/j.enconman.2022.116635.

[13] Howells CA. Piezoelectric energy harvesting. *Energy Convers Manag* 2009;50:1847–50. https://doi.org/10.1016/j.enconman.2009.02.020.

[14] Yang Z, Zhou S, Zu J, Inman D. High-performance piezoelectric energy harvesters and their applications. *Joule* 2018;2:642–97. https://doi.org/10.1016/j.joule.2018.03.011.

[15] Yang Z, Zu J. High-efficiency compressive-mode energy harvester enhanced by a multi-stage force amplification mechanism. *Energy Convers Manag* 2014;88:829–33. https://doi.org/10.1016/j.enconman.2014.09.026.

[16] Marzencki M, Basrour S, Charlot B, Grasso A, Colin M, Valbin L. Design and fabrication of piezoelectric micro power generators for autonomous microsystems. *Symp Des Test Integr Packag MEMS/MOEMS (DTIP'05)*, June 1–3, Montreux, Switzerland: TIMA; 2005, pp. 299–302.

[17] Kumar A, Balpande SS, Anjankar SC. Electromagnetic energy harvester for low frequency vibrations using MEMS. *Procedia Comput Sci* 2016;79:785–92. https://doi.org/10.1016/j.procs.2016.03.104.

[18] Beeby SP, O'Donnell T. Electromagnetic energy harvesting. 2009. https://doi.org/10.1007/978-0-387-76464-1_5.

[19] Bakhtiar S, Khan FU. Analytical modeling and simulation of an electromagnetic energy harvester for pulsating fluid flow in pipeline. *Sci World J* 2019;2019. https://doi.org/10.1155/2019/5682517.

[20] Khan FU, Iqbal M, Khan FU and Iqbal M. Electromagnetic bridge energy harvester utilizing bridge's vibrations and ambient wind for wireless sensor node application. *J Sens* 2018;2018:18. https://doi.org/10.1155/2018/3849683.

[21] Aljadiri RT, Taha LY, Ivey P. Electrostatic energy harvesting systems: A better understanding of their sustainability electrostatic energy harvesting systems. *J Clean Energy Technol* 2017;5:409–16. https://doi.org/10.18178/jocet.2017.5.5.407.

[22] Ahmad MR, Md Khir MH, Dennis JO, Zain AM. Fabrication and characterization of the electrets material for electrostatic energy harvester. *J Phys Conf Ser* 2013;476:1–6. https://doi.org/10.1088/1742-6596/476/1/012120.

[23] Crovetto A, Wang F, Hansen O. Modeling and optimization of an electrostatic energy harvesting device. *J Microelectromech Syst* 2014;23:1141–55. https://doi.org/10.1109/JMEMS.2014.2306963.

[24] Boisseau S, Despesse G, Seddik BA. Electrostatic conversion for vibration energy harvesting. *Small-Scale Energy Harvest* 2012:1–39. https://doi.org/10.5772/51360.

[25] Fu Y, Ouyang H, Davis RB. Triboelectric energy harvesting from the vibro-impact of three cantilevered beams. *Mech Syst Signal Process* 2019;121:509–31. https://doi.org/10.1016/j.ymssp.2018.11.043.

[26] Sano C, Mitsuya H, Ono S, Miwa K, Toshiyoshi H, Fujita H. Triboelectric energy harvesting with surface-charge-fixed polymer based on ionic liquid.

*Sci Technol Adv Mater* 2018;19:317–23. https://doi.org/10.1080/14686996.2018.1448200.

[27] Berry AL. The application of a triboelectric energy harvester in the packaged product vibration. *Environment.* 2016.

[28] Lu W, Xu Y, Zou Y, *et al.* Corrosion-resistant and high-performance crumpled-platinum-based triboelectric nanogenerator for self-powered motion sensing. *Nano Energy* 2020;69:104430. https://doi.org/10.1016/j.nanoen.2019.104430.

[29] Guo L, Lu Q. Potentials of piezoelectric and thermoelectric technologies for harvesting energy from pavements. *Renew Sustain Energy Rev* 2017;72:761–73. https://doi.org/10.1016/j.rser.2017.01.090.

[30] Khan F, Sassani F, Stoeber B. Copper foil-type vibration-based electromagnetic energy harvester. *J Micromech Microeng* 2010;20:125006. https://doi.org/10.1088/0960-1317/20/12/125006.

[31] Firouzi B, Abbasi A, Sendur P, Zamanian M, Chen H. Enhancing the performance of piezoelectric energy harvester under electrostatic actuation using a robust metaheuristic algorithm. *Eng Appl Artif Intell* 2023;118:105619. https://doi.org/https://doi.org/10.1016/j.engappai.2022.105619.

[32] Shi Q, Qiu C, He T, *et al.* Triboelectric single-electrode-output control interface using patterned grid electrode. *Nano Energy* 2019;60:545–56. https://doi.org/10.1016/j.nanoen.2019.03.090.

[33] Ahmed A, Hassan I, Mosa IM, *et al.* All printable snow-based triboelectric nanogenerator. *Nano Energy* 2019;60:17–25. https://doi.org/10.1016/j.nanoen.2019.03.032.

[34] Zhang R, Hummelgård M, Örtegren J, Olsen M, Andersson H, Olin H. Interaction of the human body with triboelectric nanogenerators. *Nano Energy* 2019;57:279–92. https://doi.org/10.1016/j.nanoen.2018.12.059.

[35] Parida K, Xiong J, Zhou X, Lee PS. Progress on triboelectric nanogenerator with stretchability, self-healability and bio-compatibility. *Nano Energy* 2019;59:237–57. https://doi.org/10.1016/j.nanoen.2019.01.077.

[36] Lai SN, Chang CK, Yang CS, *et al.* Ultrasensitivity of self-powered wireless triboelectric vibration sensor for operating in underwater environment based on surface functionalization of rice husks. *Nano Energy* 2019;60:715–23. https://doi.org/10.1016/j.nanoen.2019.03.067.

[37] Ravichandran AN, Calmes C, Serres JR, Ramuz M, Blayac S. Compact and high performance wind actuated venturi triboelectric energy harvester. *Nano Energy* 2019. https://doi.org/10.1016/j.nanoen.2019.05.053.

[38] Dharmasena RDIG, Silva SRP. Towards optimized triboelectric nanogenerators. *Nano Energy* 2019. https://doi.org/10.1016/j.nanoen.2019.05.057.

[39] Zhao G, Zhang Y, Shi N, *et al.* Transparent and stretchable triboelectric nanogenerator for self-powered tactile sensing. *Nano Energy* 2019;59:302–10. https://doi.org/10.1016/j.nanoen.2019.02.054.

[40] Zuo L, Tang X. Large-scale vibration energy harvesting. *J Intell Mater Syst Struct* 2013;24:1405–30. https://doi.org/10.1177/1045389X13486707.

[41] Min CL, Li Z, Xu LZ, Guo C, Peng LP, Qiang MX, Ming WT. Design and test of the MEMS coupled piezoelectric–electromagnetic energy harvester. *Int J Precis Eng Manuf* 2019;20:673–86. https://doi.org/10.1007/s12541-019-00051-x.

[42] Fan K, Liu S, Liu H, Zhu Y, Wang W, Zhang D. Scavenging energy from ultra-low frequency mechanical excitations through a bi-directional hybrid energy harvester. *Appl Energy* 2018;216:8–20. https://doi.org/10.1016/j.apenergy.2018.02.086.

[43] Fan K, Tan Q, Liu H, Zhu Y, Wang W, Zhang D. Hybrid piezoelectric-electromagnetic energy harvester for scavenging energy from low-frequency excitations. *Smart Mater Struct* 2018;27:1–46. https://doi.org/10.1088/1361-665X/aaae92.

[44] Aamir M, Waqas M, Iqbal M, Imran Hanif M, Muhammad R. Fuzzy logic approach for investigation of microstructure and mechanical properties of Sn96.5-Ag3.0-Cu0.5 lead free solder alloy. *Solder Surf Mt Technol* 2017;29:191–8. https://doi.org/10.1108/SSMT-02-2017-0005.

[45] He J, Wen T, Qian S, *et al.* Triboelectric-piezoelectric-electromagnetic hybrid nanogenerator for high-efficient vibration energy harvesting and self-powered wireless monitoring system. *Nano Energy* 2018;43:326–39. https://doi.org/10.1016/j.nanoen.2017.11.039.

[46] O'Connor SM, Lynch JP, Gilbert AC. Compressed sensing embedded in an operational wireless sensor network to achieve energy efficiency in long-term monitoring applications. *Smart Mater Struct* 2014;23. https://doi.org/10.1088/0964-1726/23/8/085014.

[47] Parkkila T, Leinonen J, Leinonen P, Oyj R. Wireless communication and mems sensors for cheaper condition monitoring and prognostics of charging crane. *Proc 19th Int Congr, Condition Monitoring and Diagnostic Engineering Managment*; 2006, pp. 747–55.

[48] Lee D. Wireless and powerless sensing node system developed for monitoring motors. *Sensors* 2008;8:5005–22. https://doi.org/10.3390/s8085005.

[49] Jovanov E, Milenkovic A, Otto C, De Groen PC. A wireless body area network of intelligent motion sensors for computer assisted physical rehabilitation. *J Neuroeng Rehabil* 2005;2:1–10. https://doi.org/10.1186/1743-0003-2-6.

[50] D'Auteuil S, Birjandi A, Bibeau E, Jordan S, Soviak J, Friesen D. Riverine hydrokinetic resource assessment using low cost winter imagery. *Renew Sustain Energy Rev* 2019;105:293–300. https://doi.org/10.1016/j.rser.2019.01.057.

[51] Ayinla SL, Inyiama HC, Azubogu ACO, Dilibe CG. Empirical analysis of energy consumption of acceleration-based wireless sensor nodes used for pipeline monitoring. *Int J Innov Res Sci Eng Technol* 2017;6:10144–9. https://doi.org/10.15680/IJIRSET.2017.0606004.

[52] Kurata N, Suzuki M, Saruwatari S, *et al.* Actual application of ubiquitous structural monitoring system using wireless sensor networks. *14th World Conference on Earthquake Engineering*, Beijing, China, October 12–17, 2008.

[53] Chen B, Wang J. Design of a multi-modal and high computation power wireless sensor node for structural health monitoring. *2008 IEEE/ASME Int*

*Conf Mechatron Embed Syst Appl MESA* 2008 2008:420–5. https://doi.org/10.1109/MESA.2008.4735683.

[54] Kurata N, Saruwatarit S, Morikawat H. Ubiquitous structural monitoring using wireless sensor networks. *Int Symp Intell Signal Process Commun Syst* 2006:99–102.

[55] Rice JA, Spencer B. Flexible smart sensor framework for autonomous full-scale structural health monitoring. *NSEL Rep* 2009;6:423–38. https://doi.org/10.12989/sss.2010.6.5_6.423.

[56] Pakzad SN, Fenves GL, Kim S, Culler DE. Design and implementation of scalable wireless sensor network for structural monitoring. *J Infrastruct Syst* 2008;14:89–101. https://doi.org/10.1061/(ASCE)1076-0342(2008)14:1(89).

[57] Dunn B, Dunn B, Kamath H, Tarascon J. Electrical energy storage for the gridfor the grid: A battery of choices. *Sci Mag* 2011;334:928–36. https://doi.org/10.1126/science.1212741.

[58] Cook-Chennault KA, Thambi N, Sastry AM. Powering MEMS portable devices – A review of non-regenerative and regenerative power supply systems with special emphasis on piezoelectric energy harvesting systems. *Smart Mater Struct* 2008;17:043001. https://doi.org/10.1088/0964-1726/17/4/043001.

[59] Aghamohammadi MR, Abdolahinia H. A new approach for optimal sizing of battery energy storage system for primary frequency control of islanded microgrid. *Int J Electr Power Energy Syst* 2014;54:325–33. https://doi.org/10.1016/j.ijepes.2013.07.005.

[60] Koo M, Park K Il, Lee SH, *et al.* Bendable inorganic thin-film battery for fully flexible electronic systems. *Nano Lett* 2012;12:4810–6. https://doi.org/10.1021/nl302254v.

[61] Zhu J, Zhang X, Luo H, Sahraei E. Investigation of the deformation mechanisms of lithium-ion battery components using in-situ micro tests. *Appl Energy* 2018;224:251–66. https://doi.org/10.1016/j.apenergy.2018.05.007.

[62] Alahmad MA, Hess HL. Evaluation and analysis of a new solid-state rechargeable microscale lithium battery. *IEEE Trans Ind Electron* 2008;55:3391–401. https://doi.org/10.1109/TIE.2008.925318.

[63] Vullers RJM, van Schaijk R, Doms I, Van Hoof C, Mertens R. Micropower energy harvesting. *Solid State Electron* 2009;53:684–93. https://doi.org/10.1016/j.sse.2008.12.011.

[64] Wang L, Yu J, Dong X, *et al.* Three-dimensional macroporous carbon/$Fe_3O_4$-doped porous carbon nanorods for high-performance supercapacitor. *ACS Sustain Chem Eng* 2016;4:1531–7. https://doi.org/10.1021/acssuschemeng.5b01474.

[65] Veneri O, Capasso C, Patalano S. Experimental investigation into the effectiveness of a super-capacitor based hybrid energy storage system for urban commercial vehicles. *Appl Energy* 2017;227:312–23. https://doi.org/10.1016/j.apenergy.2017.08.086.

[66] Zhu WH, Tatarchuk BJ. Characterization of asymmetric ultracapacitors as hybrid pulse power devices for efficient energy storage and power delivery

applications. *Appl Energy* 2016;169:460–8. https://doi.org/10.1016/j.apenergy.2016.02.020.

[67] Aneke M, Wang M. Energy storage technologies and real life applications – A state of the art review. *Appl Energy* 2016;179:350–77. https://doi.org/10.1016/j.apenergy.2016.06.097.

[68] Ma T, Yang H, Lu L. Development of hybrid battery-supercapacitor energy storage for remote area renewable energy systems. *Appl Energy* 2015;153:56–62. https://doi.org/10.1016/j.apenergy.2014.12.008.

[69] Lin Z, Wu Y, He Q, *et al.* An airtight-cavity-structural triboelectric nanogenerator-based insole for high performance biomechanical energy harvesting. *Nanoscale* 2019;11:6802–9. https://doi.org/10.1039/c9nr00083f.

[70] He W, Wang J. Optimal selection of air expansion machine in compressed air energy storage: A review. *Renew Sustain Energy Rev* 2018;87:77–95. https://doi.org/10.1016/j.rser.2018.01.013.

[71] Hannan MA, Hoque MM, Mohamed A, Ayob A. Review of energy storage systems for electric vehicle applications: Issues and challenges. *Renew Sustain Energy Rev* 2017;69:771–89. https://doi.org/10.1016/j.rser.2016.11.171.

[72] Chen J, Qiu Q, Han Y, Lau D. Piezoelectric materials for sustainable building structures: Fundamentals and applications. *Renew Sustain Energy Rev* 2019;101:14–25. https://doi.org/10.1016/j.rser.2018.09.038.

[73] Aditya L, Mahlia TMI, Rismanchi B, *et al.* A review on insulation materials for energy conservation in buildings. *Renew Sustain Energy Rev* 2017;73:1352–65. https://doi.org/10.1016/j.rser.2017.02.034.

[74] Lehmann M, Karimpour F, Goudey CA, Jacobson PT, Alam MR. Ocean wave energy in the United States: Current status and future perspectives. *Renew Sustain Energy Rev* 2017;74:1300–13. https://doi.org/10.1016/j.rser.2016.11.101.

[75] Li M, Jing X. Novel tunable broadband piezoelectric harvesters for ultralow-frequency bridge vibration energy harvesting. *Appl Energy* 2019; 255:113829. https://doi.org/10.1016/j.apenergy.2019.113829.

[76] Zhou M, Al-Furjan MSH, Zou J, Liu W. A review on heat and mechanical energy harvesting from human: Principles, prototypes and perspectives. *Renew Sustain Energy Rev* 2018;82:3582–609. https://doi.org/10.1016/j.rser.2017.10.102.

[77] Roundy S, Wright PK, Rabaey J. A study of low level vibrations as a power source for wireless sensor nodes. *Comput Commun* 2003;26:1131–44. https://doi.org/10.1016/S0140-3664(02)00248-7.

[78] Kim Y, Keun H. Elastic member and vibration absorption apparatus for a refrigerator compressor. US6,912,865 B2, 2005.

[79] Leland ES, Lai EM, Wright PK. A self-powered wireless sensor for indoor environmental monitoring. *Wirel Netw Symp* 2004, pp. 1–5.

[80] Torah R, Glynne-Jones P, Tudor M, O'Donnell T, Roy S, Beeby S. Self-powered autonomous wireless sensor node using vibration energy harvesting. *Meas Sci Technol* 2008;19:125202. https://doi.org/10.1088/0957-0233/19/12/125202.

[81] Cook-Chennault KA, Thambi N, Bitetto MA, Hameyie EB. Piezoelectric energy harvesting. *Bull Sci Technol Soc* 2008;28:496–509. https://doi.org/10.1177/0270467608325374.

[82] Roundy S. On the effectiveness of vibration-based energy harvesting. *J Intell Mater Syst Struct* 2005;16:809–23. https://doi.org/10.1177/1045389X05054042.

[83] Spelta C, Previdi F, Savaresi SM, Fraternale G, Gaudiano N. Control of magnetorheological dampers for vibration reduction in a washing machine. *Mechatronics* 2009;19:410–21. https://doi.org/10.1016/j.mechatronics.2008.09.006.

[84] Priya S. Advances in energy harvesting using low profile piezoelectric transducers. *J Electroceramics* 2007;19:165–82. https://doi.org/10.1007/s10832-007-9043-4.

[85] Kehoe MW, Freudinger LC. Aircraft ground vibration testing at the NASA Dryden Flight Research Facility – 1993 aircraft ground vibration testing at the NASA Dryden Flight Research Facility – 1993 1994.

[86] Yoon YJ, Park WT, Li KHH, Ng YQ, Song Y. A study of piezoelectric harvesters for low-level vibrations in wireless sensor networks. *Int J Precis Eng Manuf* 2013;14:1257–62. https://doi.org/10.1007/s12541-013-0171-2.

[87] M'Boungui G, Adendorff K, Naidoo R, Jimoh AA, Okojie DE. A hybrid piezoelectric micro-power generator for use in low power applications. *Renew Sustain Energy Rev* 2015;49:1136–44. https://doi.org/10.1016/j.rser.2015.04.143.

[88] Ji SH, Cho YS, Yun JS. Wearable core–shell piezoelectric nanofiber yarns for body movement energy harvesting. *Nanomaterials* 2019;9:1–9. https://doi.org/10.3390/nano9040555.

[89] Lee J, Choi B. Development of a piezoelectric energy harvesting system for implementing wireless sensors on the tires. *Energy Convers Manag* 2014;78:32–8. https://doi.org/10.1016/j.enconman.2013.09.054.

[90] Kyono T, Suzuki RO, Ono K. Conversion of unused heat energy to electricity by means of thermoelectric generation in condenser. *IEEE Trans Energy Convers* 2003;18:330–4. https://doi.org/10.1109/TEC.2003.811721.

[91] Ackermann T, Soder L. Wind energy technology and current status: a review. *Renew Sustain Energy Rev* 2000;4:315–74. https://doi.org/10.1016/S1364-0321(00)00004-6.

[92] Farhangdoust S, Mehrabi A, Younesian D. Bistable wind-induced vibration energy harvester for self-powered wireless sensors in smart bridge monitoring systems. *Proc SPIE* 2019;10971.

[93] Farhat M, Barambones O, Sbita L. A new maximum power point method based on a sliding mode approach for solar energy harvesting. *Appl Energy* 2017;185:1185–98. https://doi.org/10.1016/j.apenergy.2016.03.055.

[94] Kimura S, Sugou T, Tomioka S, Iizumi S, Tsujimoto K, Yasushiro N. Acoustic energy harvester fabricated using sol/gel lead zirconate titanate thin film. *Jpn J Appl Phys* 2011;50:187–90.

[95] Reimers CE, Tender LM, Fertig S, Wang W. Harvesting energy from the marine sediment–water interface. *Environ Sci Technol* 2001;35:192–5. https://doi.org/10.1016/j.jpowsour.2008.06.079.

[96] Ahmed A, Abu Bakar MS, Azad AK, Sukri RS, Mahlia TMI. Potential thermochemical conversion of bioenergy from *Acacia* species in Brunei Darussalam: A review. *Renew Sustain Energy Rev* 2018;82:3060–76. https://doi.org/10.1016/j.rser.2017.10.032.

[97] Hossain N, Zaini J, Meurah T, Mahlia I. Life cycle assessment, energy balance and sensitivity analysis of bioethanol production from microalgae in a tropical country. *Renew Sustain Energy Rev* 2019;115:109371. https://doi.org/10.1016/j.rser.2019.109371.

[98] Rabaey JMM, Ammer MJJ, Silva JLJ Da, *et al.* PicoRadio supports ad hoc ultra-low power wireless networking. *Computer (Long Beach Calif)* 2000;33:42–8. https://doi.org/10.1109/2.869369.

[99] Hande A, Polk T, Walker W, Bhatia D. Indoor solar energy harvesting for sensor network router nodes. *Microprocess Microsyst* 2007;31:420–32. https://doi.org/10.1016/j.micpro.2007.02.006.

[100] Ringeisen BR, Henderson E, Wu PK, *et al.* High power density from a miniature microbial fuel cell using *Shewanella oneidensis* DSP10. *Environ Sci Technol* 2006;40:2629–34. https://doi.org/10.1021/es052254w.

[101] Nayyar A, Stoilov V. Power generation from airflow induced vibrations. *Wind Eng* 2015;39:175–82. https://doi.org/10.1260/0309-524X.39.2.175.

[102] Wang C, Li Y, Chui YS, Wu QH, Chen X, Zhang W. Three-dimensional Sn-graphene anode for high-performance lithium-ion batteries. *Nanoscale* 2013;5:10599–604. https://doi.org/10.1039/c3nr02872k.

[103] Li Y, Liu K, Foley AM, Zülke A, *et al.* Data-driven health estimation and lifetime prediction of lithium-ion batteries: A review. *Renew Sustain Energy Rev* 2019;113. https://doi.org/10.1016/j.rser.2019.109254.

[104] Horowitz SB, Sheplak M, Cattafesta LN, Nishida T. A MEMS acoustic energy harvester. *J Micromech Microeng* 2006;16:S174–S181. https://doi.org/10.1088/0960-1317/16/9/S02.

[105] Cahill P, Hazra B, Karoumi R, Mathewson A, Pakrashi V. Vibration energy harvesting based monitoring of an operational bridge undergoing forced vibration and train passage. *Mech Syst Signal Process* 2018;106:265–283. https://doi.org/10.1016/j.ymssp.2018.01.007.

[106] Raghunathan V, Kansal A, Hsu J, Friedman J, Srivastava M. Design considerations for solar energy harvesting wireless embedded systems. *Proc 4th Int Symp Inf Process Sens Networks* 2005;00:511. https://doi.org/10.1109/IPSN.2005.1440973.

[107] Callebaut G, Leenders G, Van Mulders J, Ottoy G, De Strycker L, Van der Perre L. The art of designing remote IoT devices: Technologies and strategies for a long battery life. *Sensors* 2021;21:913.

[108] Maamer B, Boughamoura A, El-bab AMRF, Francis LA. A review on design improvements and techniques for mechanical energy harvesting using piezoelectric and electromagnetic schemes. *Energy Convers Manag* 2019;199:111973. https://doi.org/10.1016/j.enconman.2019.111973.

[109] Sodano HA, Inman DJ, Park G. A review of power harvesting from vibration using piezoelectric materials. *Shock Vib Dig* 2004;36:197–205. https://doi.org/10.1177/0583102404043275.

[110] Fang D, Liu J. Basic equations of piezoelectric materials. *Fracture Mechanics of Piezoelectric and Ferroelectric Solids*, Berlin: Springer; 2013, pp. 77–95. https://doi.org/10.1007/978-3-642-30087-5_4.

[111] Triplett A, Quinn DD. The effect of non-linear piezoelectric coupling on vibration-based energy harvesting. *J Intell Mater Syst Struct* 2009; 20: 1959–67. https://doi.org/10.1177/1045389X09343218.

[112] Wang H, Jasim A, Chen X. Energy harvesting technologies in roadway and bridge for different applications: A comprehensive review. *Appl Energy* 2018;212:1083–94. https://doi.org/10.1016/j.apenergy.2017.12.125.

[113] Wang C, Zhao J, Li Q, Li Y. Optimization design and experimental investigation of piezoelectric energy harvesting devices for pavement. *Appl Energy* 2018;229:18–30. https://doi.org/10.1016/j.apenergy.2018.07.036.

[114] Orfei F, Vocca H, Gammaitoni L. Linear and non-linear energy harvesting from bridge vibrations. *Proc ASME 2016 Int Des Eng Tech Conf Comput Inf Eng Conf*, 2016, pp. 1–8. https://doi.org/10.3945/jn.108.096511.

[115] Wu W, Lee B, Jong W, Shiun B. Piezoelectric MEMS power generators for vibration energy harvesting. *Small-Scale Energy Harvest* 2012. https://doi.org/10.5772/51997.

[116] Kurt E, Mustafa MK, Akbaba S, Bizon N. Analytical and experimental studies on a new linear energy harvester. *Can J Phys* 2018;96:727–33. https://doi.org/10.1139/cjp-2017-0708.

[117] Halvorsen E. Fundamental issues in nonlinear wideband-vibration energy harvesting. *Phys Rev E: Stat Nonlinear, Soft Matter Phys* 2013;87:1–6. https://doi.org/10.1103/PhysRevE.87.042129.

[118] Stanton SC, Erturk A, Mann BP, Inman DJ. Nonlinear piezoelectricity in electroelastic energy harvesters: Modeling and experimental identification. *J Appl Phys* 2010;108:1–9. https://doi.org/10.1063/1.3486519.

[119] Gammaitoni L, Vocca H, Neri I, Travasso F, Orfei F, Perugia U. Vibration energy harvesting: Linear and nonlinear oscillator approaches. *Sustain Energy Harvesting Technologies: Past, Present, and Future* 2011:885–90. https://doi.org/10.5772/25623.

[120] Stanton SC, McGehee CC, Mann BP. Nonlinear dynamics for broadband energy harvesting: Investigation of a bistable piezoelectric inertial generator. *Phys D: Nonlinear Phenom* 2010;239:640–53. https://doi.org/10.1016/j.physd.2010.01.019.

[121] Marzencki M, Basrour S, Charlot B, *et al.* A MEMS piezoelectric vibration energy harvesting device. *PowerMEMS 2005* 2005:45–8.

[122] Wei S, Hu H, He S. Modeling and experimental investigation of an impact-driven piezoelectric energy harvester from human motion. *Smart Mater Struct* 2013;22:105020. https://doi.org/10.1088/0964-1726/22/10/105020.

[123] Liu H, Lee C, Kobayashi T, Tay CJ, Quan C. Piezoelectric MEMS-based wideband energy harvesting systems using a frequency-up-conversion cantilever stopper. *Sensors Actuators A Phys* 2012;186: 242–8. https://doi.org/10.1016/j.sna.2012.01.033.

[124] Jeong SY, Hwang WS, Cho JY, *et al.* Piezoelectric device operating as sensor and harvester to drive switching circuit in LED shoes. *Energy* 2019;177:87–93. https://doi.org/10.1016/j.energy.2019.04.061.

[125] Isarakorn D, Briand D, Janphuang P, *et al.* The realization and performance of vibration energy harvesting MEMS devices based on an epitaxial piezoelectric thin film. *Smart Mater Struct* 2011;20. https://doi.org/10.1088/0964-1726/20/2/025015.

[126] Lu F, Lee HP, Lim SP. Modeling and analysis of micro piezoelectric power generators for micro-electromechanical-systems applications. *Smart Mater Struct* 2003;13:57–63. https://doi.org/10.1088/0964-1726/13/1/007.

[127] Pozzi M, Zhu M. Plucked piezoelectric bimorphs for energy harvesting. In: Elvin, N. and Erturk, A. (eds), *Advances in Energy Harvesting Methods.* Springer, New York, NY; 2013.

[128] Roundy S, Wright PK. A piezoelectric vibration based generator for wireless electronics. *Smart Mater Struct* 2004;13:1131–42. https://doi.org/10.1088/0964-1726/13/5/018.

[129] Ethem Erkan A, Hanseup K, Najafi K. Energy scavenging from insect flight. *J Micromech Microeng* 2011;21:95016. https://doi.org/doi:10.1088/0960-1317/21/9/095016.

[130] Peigney M, Siegert D. Piezoelectric energy harvesting from traffic-induced bridge vibrations. *Smart Mater Struct* 2013;22:095019. https://doi.org/10.1088/0964-1726/22/9/095019.

[131] Tang X, Wang X, Cattley R, Gu F, Ball A. Energy harvesting technologies for achieving self-powered wireless sensor networks in machine condition monitoring: A review. *Sensors* 2018;18:4113. https://doi.org/10.3390/s18124113.

[132] Barman S Das, Reza AW, Kumar N, Karim ME, Munir AB. Wireless powering by magnetic resonant coupling: Recent trends in wireless power transfer system and its applications. *Renew Sustain Energy Rev* 2015;51:1525–52. https://doi.org/10.1016/j.rser.2015.07.031.

[133] Galchev TV, McCullagh J, Peterson RL, Najafi K. Harvesting traffic-induced vibrations for structural health monitoring of bridges. *J Micromech Microeng* 2011;21:104005. https://doi.org/10.1088/0960-1317/21/10/104005.

[134] Pan Y, Lin T, Qian F, *et al.* Modeling and field-test of a compact electromagnetic energy harvester for railroad transportation. Modeling and field-test of a compact electromagnetic energy harvester for railroad transportation. *Appl Energy* 2019;247:309–21. https://doi.org/10.1016/j.apenergy.2019.03.051.

[135] Khan FU, Iqbal M. Electromagnetic bridge energy harvester utilizing bridge's vibrations and ambient wind for wireless sensor node application. *J Sens* 2018;2018:18. https://doi.org/10.1155/2018/3849683.

[136] Galchev T, Kim H, Najafi K. Non-resonant bi-stable frequency-increased power scavenger from low-frequency ambient vibration. *TRANSDUCERS 2009–2009 Int. Solid-State Sensors, Actuators Microsystems Conf*, Denver, CO: IEEE; 2009, pp. 632–5. https://doi.org/10.1109/SENSOR.2009.5285404.

[137] Zhang LB, Dai HL, Yang YW, Wang L. Design of high-efficiency electromagnetic energy harvester based on a rolling magnet. *Energy Convers Manag* 2019;185:202–10. https://doi.org/10.1016/j.enconman.2019.01.089.

[138] Wang W, Cao J, Zhang N, Lin J, Liao WH. Magnetic-spring based energy harvesting from human motions: Design, modeling and experiments. *Energy Convers Manag* 2017;132:189–97. https://doi.org/10.1016/j.enconman.2016.11.026.

[139] von Büren T. Body-worn inertial electromagnetic micro-generators. *Swiss Federal Institute of Technology Zurich*, 2006.

[140] Thein CK, Foong FM, Ooi B. Effect of repulsive magnetic poles on the natural frequency and the bandwidth of a vibration energy harvester. *J Phys Conf Ser* 2018;1123:12019. https://doi.org/10.1088/1742-6596/1123/1/012019.

[141] Aouali K, Kacem N, Mrabet E Bouhaddi N, Haddar M. Effect of the localization on the response of a quasi-periodic electromagnetic oscillator array for vibration energy harvesting. *MATEC Web Conf* 2018;241:1003. https://doi.org/10.1051/matecconf/201824101003.

[142] Ding J, Su Y, Zhang K. Structure analysis and output performance of vibration energy harvester based on diamagnetic levitation. *10th Int. Conf. Appl. Energy (ICAE2018)*, 22–25 August 2018, Hong Kong, China, Hong Kong: *Energy Procedia*; 2018;158:5575–80. https://doi.org/10.1016/j.egypro.2019.01.584.

[143] Kirolos A, Moustafa A, El-Badawy A, *et al.* Design and development of an electromagnetic micropower generator for activity tracking. *2018 20th Int Middle East Power Syst Conf MEPCON 2018 – Proc*, Piscataway, NJ: IEEE; 2019, pp. 522–7. https://doi.org/10.1109/MEPCON.2018.8635256.

[144] Yang X, Zhang B, Li J, Wang Y. Model and experimental research on an electromagnetic vibration-powered generator with annular permanent magnet spring. *IEEE Trans Appl Supercond* 2012;22:0–3. https://doi.org/10.1109/TASC.2011.2179401.

[145] Zhang Q, Zhang Z, Liang Q, *et al.* Green hybrid power system based on triboelectric nanogenerator for wearable/portable electronics. *Nano Energy* 2019;55:151–63. https://doi.org/10.1016/j.nanoen.2018.10.078.

[146] Kang H, Kim HT, Woo HJ, *et al.* Metal nanowire–polymer matrix hybrid layer for triboelectric nanogenerator. *Nano Energy* 2019;58:227–33. https://doi.org/10.1016/j.nanoen.2019.01.046.

[147] Kil TH, Kim S, Jeong DH, *et al.* A highly-efficient, concentrating-photovoltaic/thermoelectric hybrid generator. *Nano Energy* 2017;37:242–7. https://doi.org/10.1016/j.nanoen.2017.05.023.

[148] Singh HH, Khare N. Flexible ZnO–PVDF/PTFE based piezo-tribo hybrid nanogenerator. *Nano Energy* 2018;51:216–22. https://doi.org/10.1016/j.nanoen.2018.06.055.

[149] Chowdhury AR, Abdullah AM, Hussain I, *et al.* Lithium doped zinc oxide based flexible piezoelectric-triboelectric hybrid nanogenerator. *Nano Energy* 2019;61:327–36. https://doi.org/10.1016/j.nanoen.2019.04.085.

[150] Yang H, Wang M, Deng M, *et al.* A full-packaged rolling triboelectric-electromagnetic hybrid nanogenerator for energy harvesting and building up self-powered wireless systems. *Nano Energy* 2019;56:300–6. https://doi.org/10.1016/j.nanoen.2018.11.043.

[151] Wacharasindhu T, Kwon JW. A micromachined energy harvester from a keyboard using combined electromagnetic and piezoelectric conversion. *J Micromech Microeng* 2008;18:104016. https://doi.org/10.1088/0960-1317/18/10/104016.

[152] Zhang G, Gao S, Liu H, Zhang W. Design and performance of hybrid piezoelectric-electromagnetic energy harvester with trapezoidal beam and magnet sleeve. *J Appl Phys* 2019;125. https://doi.org/10.1063/1.5087024.

[153] Gupta RK, Shi Q, Dhakar L, Wang T, Heng CH, Lee C. Broadband energy harvester using non-linear polymer spring and electromagnetic/triboelectric hybrid mechanism. *Sci Rep* 2017;7:1–13. https://doi.org/10.1038/srep41396.

[154] Iqbal M, Khan FU. Hybrid vibration and wind energy harvesting using combined piezoelectric and electromagnetic conversion for bridge health monitoring applications. *Energy Convers Manag* 2018;172:611–8. https://doi.org/10.1016/j.enconman.2018.07.044.

[155] Rajarathinam M, Ali SF. Energy generation in a hybrid harvester under harmonic excitation. *Energy Convers Manag* 2018;155:10–9. https://doi.org/10.1016/j.enconman.2017.10.054.

[156] Yang B. Hybrid energy harvester based on piezoelectric and electromagnetic mechanisms. *J Micro/Nanolithography, MEMS, MOEMS* 2010;9:023002. https://doi.org/10.1117/1.3373516.

[157] Iqbal M, Nauman MM, Cheok QHN, *et al.* Design and modeling of a smart insole hybrid energy harvester. *7th Brunei Int Conf Eng Technol 2018 (BICET 2018)*, Bandar Seri Begawan, Brunei: Institution of Engineering and Technology; 2019, pp. 35 (4 pp.)–35 (4 pp.). https://doi.org/10.1049/cp.2018.1532.

[158] Rodrigues C, Gomes A, Ghosh A, Pereira A, Ventura J. Power-generating footwear based on a triboelectric-electromagnetic-piezoelectric hybrid nanogenerator. *Nano Energy* 2019;62:660–6. https://doi.org/10.1016/j.nanoen.2019.05.063.

[159] Zhang K, Wang X, Yang Y, Wang ZL. Hybridized electromagnetic-triboelectric nanogenerator for scavenging biomechanical energy for sustainably powering wearable electronics. *ACS Nano* 2015;9:3521–9. https://doi.org/10.1021/nn507455f.

[160] Iqbal M, Nauman MM, Khan FU, *et al.* Multimodal hybrid piezoelectric-electromagnetic insole energy harvester using PVDF generators. *Electronics* 2020;9:9040635. https://doi.org/10.3390/electronics9040635.

[161] Liu H, Gao S, Wu J, Li P. Study on the output performance of a nonlinear hybrid piezoelectric-electromagnetic harvester under harmonic excitation. *Acoustics* 2019;1:382–92. https://doi.org/10.3390/acoustics1020021.

[162] White NM, Glynne-Jones P, Beeby SP. A novel thick-film piezoelectric micro-generator. *SMART Mater Struct* 2001;10:850–2. https://doi.org/stacks.iop.org/SMS/10/850 Abstract.

[163] Jeon YB, Sood R, Jeong JH, Kim SG. MEMS power generator with transverse mode thin film PZT. *Sensors Actuators A Phys* 2005;122:16–22. https://doi.org/10.1016/j.sna.2004.12.032.

[164] Fakhzan MN, Muthalif AGA. Vibration based energy harvesting using piezoelectric material. 2011 *4th Int Conf Mechatron* 2011:1–7. https://doi.org/10.1109/ICOM.2011.5937182.

[165] Ottman GK, Hofmann HF, Bhatt AC, Lesieutre G Adaptive piezoelectric energy harvesting circuit for wireless remote power supply. *IEEE Trans Power Electron* 2002;17:669–76. https://doi.org/10.1109/TPEL.2002.802194.

[166] Khan FU, Ali T. A piezoelectric based energy harvester for simultaneous energy generation and vibration isolation. *Int J Energy Res* 2019:1–10. https://doi.org/10.1002/er.4700.

[167] Banerjee P, Banerjee P, Dhal SS. *Int J Adv Res Comput Sci Softw Eng* 2012;2:62–70.

[168] Rezaeisaray M, Gowini M El, Sameoto D, Raboud D, Moussa W. Low frequency piezoelectric energy harvesting at multi vibration mode shapes. *Sensors Actuators A Phys* 2015;228:104–11. https://doi.org/10.1016/j.sna.2015.02.036.

[169] Liu D, Li H, Feng H, Yalkun T, Hajj MR. A multi-frequency piezoelectric vibration energy harvester with liquid filled container as the proof mass. *Appl Phys Lett* 2019;114. https://doi.org/10.1063/1.5089289.

[170] Tang Q, Li X. Two-stage wideband energy harvester driven by multimode coupled vibration. *IEEE/ASME Trans Mechatron* 2015;20:115–21. https://doi.org/10.1109/TMECH.2013.2296776.

[171] Jasim A, Wang H, Yesner G, Safari A, Maher A. Optimized design of layered bridge transducer for piezoelectric energy harvesting from roadway. *Energy* 2017;141:1133–45. https://doi.org/10.1016/j.energy.2017.10.005.

[172] Jasim A, Yesner G, Wang H, Safari A, Maher A, Basily B. Laboratory testing and numerical simulation of piezoelectric energy harvester for roadway applications. *Appl Energy* 2018;224:438–47. https://doi.org/10.1016/j.apenergy.2018.05.040.

[173] Liu H, Tay CJ, Quan C, Kobayashi T, Lee C. Piezoelectric MEMS energy harvester for low-frequency vibrations with wideband operation range and steadily increased output power. *J Microelectromech Syst* 2011;20:1131–42. https://doi.org/10.1109/JMEMS.2011.2162488.

[174] Bakhtiar S, Khan FU, Rahman WU, Khan AS, Ahmad MM, Iqbal M. A pressure-based electromagnetic energy harvester for pipeline monitoring applications. *J Sens* 2022;2022:6529623. https://doi.org/10.1155/2022/6529623.

[175] Galchev T, McCullagh J, Peterson RL, Najafi K. Harvesting traffic-induced bridge vibrations. *2011 16th Int Solid-State Sensors, Actuators Microsystems Conf TRANSDUCERS'11*, 2011, pp. 1661–4. https://doi.org/10.1109/TRANSDUCERS.2011.5969860.

[176] Marzencki M, Basrour S, Charlot B. Design, modelling and optimisation of integrated piezoelectric micro power generators – The method design optimization. *Simulation* 2005;3:545–8.

[177] Galchev T, Kim H, Najafi K. Micro power generator for harvesting low-frequency and nonperiodic vibrations. *Microelectromech Syst* 2011; 20:852–66.

[178] Khan FU, Ahmad I. Vibration-based electromagnetic type energy harvester for bridge monitoring sensor application. *Proc. – 2014 Int. Conf. Emerg. Technol. ICET 2014*, 2014, pp. 125–9. https://doi.org/10.1109/ICET.2014.7021029.

[179] Dierks EC. Design of an electromagnetic vibration energy harvester for structural health monitoring of bridges employing wireless sensor networks. The University of Texas at Austin, 2011.

[180] Khan FU, Iqbal M. Electromagnetic-based bridge energy harvester using traffic-induced bridge's vibrations and ambient wind. *2016 Int Conf Intell Syst Eng ICISE 2016*, Islamabad: IEEE; 2016, pp. 380–5. https://doi.org/10.1109/INTELSE.2016.7475152.

[181] Galchev T, McCullagh J, Peterson RL, Najafi K. A vibration harvesting system for bridge health monitoring applications. *Proc PowerMEMS 2010,* Leuven, Belgium 2010, pp. 3–6.

[182] Iqbal M, Khan FU, Mehdi M, Cheok Q, Abas E, Nauman MM. Power harvesting footwear based on piezo-electromagnetic hybrid generator for sustainable wearable microelectronics. *J King Saud Univ – Eng Sci* 2020;34:329–38. https://doi.org/10.1016/j.jksues.2020.11.003.

[183] Yu H, Zhou J, Yi X, Wu H, Wang W. A hybrid micro vibration energy harvester with power management circuit. *Microelectron Eng* 2015;131:36–42. https://doi.org/https://doi.org/10.1016/j.mee.2014.10.008.

[184] Toyabur RMM, Salauddin M, Cho H, Park JY. A multimodal hybrid energy harvester based on piezoelectric-electromagnetic mechanisms for low-frequency ambient vibrations. *Energy Convers Manag* 2018;168:454–66. https://doi.org/10.1016/j.enconman.2018.05.018.

[185] Khan FU, Iqbal M. Electromagnetic-based bridge energy harvester using traffic-induced bridge's vibrations and ambient wind. *2016 Int Conf Intell Syst Eng*, 2016, pp. 380–5. https://doi.org/10.1109/INTELSE.2016.7475152.

[186] Shenck NS, Paradiso JA. Energy scavenging with shoe-mounted piezo-electrics. *IEEE Micro* 2001;21:30–42. https://doi.org/10.1109/40.928763.

[187] Shi G, Peng Y, Tong D, *et al.* An ultra-low frequency vibration energy harvester with zigzag piezoelectric spring actuated by rolling ball. *Energy Convers Manag* 2021;243:114439. https://doi.org/10.1016/j.enconman.2021.114439.

[188] Purwadi AM, Parasuraman S, Khan MKAA, Elamvazuthi I. Development of biomechanical energy harvesting device using heel strike. *Procedia Comput Sci* 2015;76:270–5. https://doi.org/10.1016/j.procs.2015.12.288.

[189] von Büren T. Body-worn inertial electromagnetic micro-generators. *Swiss Federal Institute of Technology Zurich*, 2006.

[190] Renaud M, Fiorini P, Van Schaijk R, Van Hoof C. Harvesting energy from the motion of human limbs: The design and analysis of an impact-based piezoelectric generator. *Smart Mater Struct* 2009;18. https://doi.org/10.1088/0964-1726/18/3/035001.

[191] Charlon Y, Campo E, Brulin D. Design and evaluation of a smart insole: Application for continuous monitoring of frail people at home. *Expert Syst Appl* 2018;95:57–71. https://doi.org/10.1016/j.eswa.2017.11.024.

[192] Farid S, Khan U. Acoustic energy harvesting for autonomous wireless n.d.

[193] Khan SFU. Vibration-based electromagnetic energy harvesters for MEMS applications, University of British Columbia, 2011.

[194] Zhang Z, Xiang H, Shi Z, Zhan J. Experimental investigation on piezoelectric energy harvesting from vehicle-bridge coupling vibration. *Energy Convers Manag* 2018;163:169–79. https://doi.org/10.1016/j.enconman.2018.02.054.

[195] Halim MA, Tao K, Towfighian S, Zhu D. Vibration energy harvesting: Linear, nonlinear, and rotational approaches. *Shock Vib* 2019;2019: 5381756. https://doi.org/10.1155/2019/5381756.

[196] Hossain N, Haji Zaini J, Mahlia TMI. A review of bioethanol production from plant-based waste biomass by yeast fermentation. *Int J Technol* 2017;8:5. https://doi.org/10.14716/ijtech.v8i1.3948.

[197] Ahmed A, Hidayat S, Abu Bakar MS, Azad AK, Sukri RS, Phusunti N. Thermochemical characterisation of *Acacia auriculiformis* tree parts via proximate, ultimate, TGA, DTG, calorific value and FTIR spectroscopy analyses to evaluate their potential as a biofuel resource. *Biofuels* 2018:1–12. https://doi.org/10.1080/17597269.2018.1442663.

[198] Hossain N, Rafidah Jalil. Analysis of bio-energy properties from Malaysian local plants: Sentang and Sesendok. *Asia Pacific J Energy Environ* 2015;2:141–4. https://doi.org/10.18034/apjee.v2i3.737.

[199] Anjum MU, Fida A, Ahmad I, Iftikhar A. A broadband electromagnetic type energy harvester for smart sensor devices in biomedical applications. *Sensors Actuators A Phys* 2018;277:52–9. https://doi.org/10.1016/j.sna.2018.05.001.

[200] Thomson WWT. *Theory of Vibration with Applications*. 3rd edn. Englewood Cliffs, NJ: Prentice-Hall; 1998.

[201] Aamir M, Waqas M, Iqbal M, Imran Hanif M, Muhammad R. Fuzzy logic approach for investigation of microstructure and mechanical properties of $Sn_{96.5}$–$Ag_{3.0}$–$Cu_{0.5}$ lead free solder alloy. *Solder Surf Mt Technol* 2017;29:191–8. https://doi.org/10.1108/SSMT-02-2017-0005.

[202] Umar MM, Silva LC De, Bakar MSA, Petra MI. State of the art of smoke and fire detection using image processing. *Int J Signal Imaging Syst Eng* 2017;10:22. https://doi.org/10.1504/IJSISE.2017.084566.

[203] Cima MJ. Next-generation wearable electronics. *Nat Biotechnol* 2014;32:642–3. https://doi.org/10.1038/nbt.2952.

[204] Gao W, Emaminejad S, Nyein HYY, *et al.* Fully integrated wearable sensor arrays for multiplexed in situ perspiration analysis. *Nature* 2016;529:509–14. https://doi.org/10.1038/nature16521.

[205] Waterbury AC, Wright PK. Vibration energy harvesting to power condition monitoring sensors for industrial and manufacturing equipment. *Proc Inst Mech Eng Part C J Mech Eng Sci* 2013;227:1187–202. https://doi.org/10.1177/0954406212457895.

[206] Halim MA, Cho H, Park JY. Design and experiment of a human-limb driven, frequency up-converted electromagnetic energy harvester. *Energy*

*Convers Manag* 2015;106:393–404. https://doi.org/10.1016/j.enconman.2015.09.065.

[207] Kuang Y, Ruan T, Chew ZJ, Zhu M. Energy harvesting during human walking to power a wireless sensor node. *Sensors Actuators A Phys* 2017;254:69–77. https://doi.org/10.1016/j.sna.2016.11.035.

[208] Zhang X, Pan H, Qi L, Zhang Z, Yuan Y, Liu Y. A renewable energy harvesting system using a mechanical vibration rectifier (MVR) for railroads. *Appl Energy* 2017;204:1535–43. https://doi.org/10.1016/j.apenergy.2017.04.064.

[209] Aldawood G, Nguyen HT, Bardaweel H. High power density spring-assisted nonlinear electromagnetic vibration energy harvester for low base-accelerations. *Appl Energy* 2019;253:113546. https://doi.org/10.1016/j.apenergy.2019.113546.

[210] Niasar EHA, Dahmardeh M, Googarchin HS. Roadway piezoelectric energy harvester design considering electrical and mechanical performances. *Proc Inst Mech Eng Part C J Mech Eng Sci* 2020;234:32–48. https://doi.org/10.1177/0954406219873366.

[211] Hong GB, Su TL, Lee JD, Hsu TC, Chen HW. Energy conservation potential in Taiwanese textile industry. *Energy Policy* 2010;38:7048–53. https://doi.org/10.1016/j.enpol.2010.07.024.

[212] Zhang D, Li W, Ying Y, Zhao H, Lin Y, Bao J. Wave energy converter of inverse pendulum with double action power take off. *Proc Inst Mech Eng Part C J Mech Eng Sci* 2013;227:2416–27. https://doi.org/10.1177/0954406213475760.

[213] Wang X, Liu Y Bin, Xiao GJ, Pan CL, Feng ZH. Dual-branch reed for resonant cavity wind energy harvester with enhanced performances. *Proc Inst Mech Eng Part C J Mech Eng Sci* 2015;229:2270–80. https://doi.org/10.1177/0954406214557168.

[214] Xie L, Cai M. Increased piezoelectric energy harvesting from human footstep motion by using an amplification mechanism. *Appl Phys Lett* 2014;105:1–5. https://doi.org/10.1063/1.4897624.

[215] Chamanian S, Uluşan H, Zorlu Ö, Külah H. Wearable battery-less wireless sensor network with electromagnetic energy harvesting system. *Sensors Actuators A Phys* 2016;249:77–84. https://doi.org/10.1016/j.sna.2016.07.020.

[216] Wu S, Luk PCKK, Li C, Zhao X, Jiao Z, Shang Y. An electromagnetic wearable 3-DoF resonance human body motion energy harvester using ferrofluid as a lubricant. *Appl Energy* 2017;197:364–74. https://doi.org/10.1016/j.apenergy.2017.04.006.

[217] Turkmen AC, Celik C. Energy harvesting with the piezoelectric material integrated shoe. *Energy* 2018;150:556–64. https://doi.org/10.1016/j.energy.2017.12.159.

[218] Lu Y, Cottone F, Boisseau S, Marty F, Galayko D, Basset P. A nonlinear MEMS electrostatic kinetic energy harvester for human-powered biomedical devices. *Appl Phys Lett* 2015;107. https://doi.org/10.1063/1.4937587.

[219] Wang J, Li S, Yi F, *et al.* Sustainably powering wearable electronics solely by biomechanical energy. *Nat Commun* 2016;7:1–8. https://doi.org/10.1038/ncomms12744.

[220] Yeatman EM. Energy harvesting from motion using rotating and gyroscopic proof masses. *Proc Inst Mech Eng Part C J Mech Eng Sci* 2008;222:27–36. https://doi.org/10.1243/09544062JMES701.

[221] Payne OR, Moss SD. Modelling and development of a robust hybrid rotary-translational vibration energy harvester. *J Intell Mater Syst Struct* 2017;28:565–77. https://doi.org/10.1177/1045389X16649703.

[222] Samad FA, Karim MF, Member S. A curved electromagnetic energy harvesting system for wearable electronics. *IEEE Sens J* 2016;16:1969–74. https://doi.org/10.1109/JSEN.2015.2500603.

[223] Abed I, Kacem N, Bouhaddi N, Bouazizi ML. Multi-modal vibration energy harvesting approach based on nonlinear oscillator arrays under magnetic levitation. *Smart Mater Struct* 2016;25:25018. https://doi.org/10.1088/0964-1726/25/2/025018.

[224] Maharjan P, Bhatta T, Salauddin Rasel M, Salauddin M, Toyabur Rahman M, Park JY. High-performance cycloid inspired wearable electromagnetic energy harvester for scavenging human motion energy. *Appl Energy* 2019;256:113987. https://doi.org/10.1016/j.apenergy.2019.113987.

[225] Pillatsch P, Yeatman EM, Holmes AS. A piezoelectric frequency up-converting energy harvester with rotating proof mass for human body applications. *Sensors Actuators A Phys* 2014;206:178–85. https://doi.org/10.1016/j.sna.2013.10.003.

[226] Hwang GT, Park H, Lee JH, *et al.* Self-powered cardiac pacemaker enabled by flexible single crystalline PMN-PT piezoelectric energy harvester. *Adv Mater* 2014;26:4880–7. https://doi.org/10.1002/adma.201400562.

[227] Yang W, Chen J, Zhu G, *et al.* Harvesting energy from the natural vibration of human walking. *ACS Nano* 2013;7:11317–24. https://doi.org/10.1021/nn405175z.

[228] Jung WS, Kang MG, Moon HG, *et al.* High output piezo/triboelectric hybrid generator. *Sci Rep* 2015;5:1–6. https://doi.org/10.1038/srep09309.

[229] Khan FU, Iqbal M. Electromagnetic bridge energy harvester utilizing bridge's vibrations and ambient wind for wireless sensor node application. *J Sens* 2018;3849683. https://doi.org/10.1155/2018/3849683.

[230] Saha CR, O'Donnell T, Wang N, McCloskey P. Electromagnetic generator for harvesting energy from human motion. *Sensors Actuators A Phys* 2008;147:248–53. https://doi.org/10.1016/j.sna.2008.03.008.

[231] Halim MA, Park JY. A non-resonant, frequency up-converted electromagnetic energy harvester from human-body-induced vibration for hand-held smart system applications. *J Appl Phys* 2014;115. https://doi.org/10.1063/1.4867216.

[232] Liu H, Gudla S, Hassani FA, Heng CH, Lian Y, Lee C. Investigation of the nonlinear electromagnetic energy harvesters from hand shaking. *IEEE Sens J* 2015;15:2356–64. https://doi.org/10.1109/JSEN.2014.2375354.

[233] Halim MA, Park JY. Modeling and experiment of a handy motion driven, frequency up-converting electromagnetic energy harvester using transverse impact by spherical ball. *Sensors Actuators A Phys* 2015;229:50–8. https://doi.org/10.1016/j.sna.2015.03.024.

[234] Halim MA, Rantz R, Zhang Q, Gu L, Yang K, Roundy S. An electromagnetic rotational energy harvester using sprung eccentric rotor, driven by pseudo-walking motion. *Appl Energy* 2018;217:66–74. https://doi.org/10.1016/j.apenergy.2018.02.093.

[235] Chamanian S, Baghaee S, Ulusan H, Zorlu Ö, Külah H, Uysal-Biyikoglu E. Powering-up wireless sensor nodes utilizing rechargeable batteries and an electromagnetic vibration energy harvesting system. *Energies* 2014; 7:6323–39. https://doi.org/10.3390/en7106323.

[236] Galchev T, Kim H, Najafi K. A parametric frequency increased power generator for scavenging low frequency ambient vibrations. *Procedia Chem* 2009;1:1439–42. https://doi.org/10.1016/j.proche.2009.07.359.

[237] Iqbal M, Nauman MM, Khan FU, Abas E, Cheok Q, Aissa B. Nonlinear multi-mode electromagnetic insole energy harvester for human-powered body monitoring sensors: Design, modeling, and characterization. *Proc Inst Mech Eng Part C J Mech Eng Sci* 2021;235:6415–26. https://doi.org/10.1177/0954406221991178.

[238] Cai M, Wang J, Liao W. Self-powered smart watch and wristband enabled by embedded generator. *Appl Energy* 2020;263:114682. https://doi.org/10.1016/j.apenergy.2020.114682.

[239] Elhalwagy AM, Ghoneem MYM, Elhadidi M. Feasibility study for using piezoelectric energy harvesting floor in buildings' interior spaces. *Energy Procedia* 2017;115:114–26. https://doi.org/10.1016/j.egypro.2017.05.012.

[240] Oh T, Islam SK, To G, Mahfouz M. A low-power CMOS energy harvesting circuit for wearable sensors using piezoelectric transducers. *2017 United States Natl Comm URSI Natl Radio Sci Meet Usn NRSM 2017* 2017. https://doi.org/10.1109/USNC-URSI-NRSM.2017.7878305.

[241] Zhong X, Yang Y, Wang X, Wang ZL. Rotating-disk-based hybridized electromagnetic-triboelectric nanogenerator for scavenging biomechanical energy as a mobile power source. *Nano Energy* 2015;13:771–80. https://doi.org/10.1016/j.nanoen.2015.03.012.

[242] Margaria R. Positive and negative work performances and their efficiencies in human locomotion. *Int Z Angew Physiol* 1968;25:339–51. https://doi.org/10.1007/BF00699624.

[243] Chen JJ, Symes MD, Fan SC, *et al.* High-performance polyoxometalate-based cathode materials for rechargeable lithium-ion batteries. *Adv Mater* 2015;27:4649–54. https://doi.org/10.1002/adma.201501088.

[244] Hyland M, Hunter H, Liu J, Veety E, Vashaee D. Wearable thermoelectric generators for human body heat harvesting. *Appl Energy* 2016;182:518–24. https://doi.org/10.1016/j.apenergy.2016.08.150.

[245] Sultana A, Alam MM, Middya TR, Mandal D. A pyroelectric generator as a self-powered temperature sensor for sustainable thermal energy harvesting

from waste heat and human body heat. *Appl Energy* 2018;221:299–307. https://doi.org/10.1016/j.apenergy.2018.04.003.

[246] Fan K, Chang J, Chao F, Pedrycz W. Design and development of a multi-purpose piezoelectric energy harvester. *Energy Convers Manag* 2015; 96: 430–9. https://doi.org/10.1016/j.enconman.2015.03.014.

[247] Wang W, Cao J, Bowen CR, Zhou S, Lin J. Optimum resistance analysis and experimental verification of nonlinear piezoelectric energy harvesting from human motions. *Energy* 2017;118:221–30. https://doi.org/10.1016/j.energy.2016.12.035.

[248] Li Z, Yang Z, Naguib H, Zu J. Design and studies on a low-frequency truss-based compressive-mode piezoelectric energy harvester. *IEEE/ASME Trans Mechatron* 2018;23:2849–58. https://doi.org/10.1109/TMECH.2018.2871781.

[249] Izadgoshasb I, Lim YY, Tang L, Padilla RV, Tang ZS, Sedighi M. Improving efficiency of piezoelectric based energy harvesting from human motions using double pendulum system. *Energy Convers Manag* 2019;184:559–70. https://doi.org/10.1016/j.enconman.2019.02.001.

[250] Berdy DF, Valentino DJ, Peroulis D. Kinetic energy harvesting from human walking and running using a magnetic levitation energy harvester. *Sensors Actuators A Phys* 2015;222:262–71. https://doi.org/10.1016/j.sna.2014.12.006.

[251] Fan J, Xiong CH, Huang ZK, Wang CB, Chen W Bin. A lightweight bio-mechanical energy harvester with high power density and low metabolic cost. *Energy Convers Manag* 2019;195:641–9. https://doi.org/10.1016/j.enconman.2019.05.025.

[252] Zhang Q, Wang Y, Kim ES. Power generation from human body motion through magnet and coil arrays with magnetic spring. *J Appl Phys* 2014;115. https://doi.org/10.1063/1.4865792.

[253] Zhu G, Bai P, Chen J, Lin Wang Z. Power-generating shoe insole based on triboelectric nanogenerators for self-powered consumer electronics. *Nano Energy* 2013;2:688–92. https://doi.org/10.1016/j.nanoen.2013.08.002.

[254] Xing F, Jie Y, Cao X, Li T, Wang N. Natural triboelectric nanogenerator based on soles for harvesting low-frequency walking energy. *Nano Energy* 2017;42:138–42. https://doi.org/10.1016/j.nanoen.2017.10.029.

[255] Rasel MSU, Park JY. A sandpaper assisted micro-structured poly-dimethylsiloxane fabrication for human skin based triboelectric energy harvesting application. *Appl Energy* 2017;206:150–8. https://doi.org/10.1016/j.apenergy.2017.07.109.

[256] Beeby SP, Torah RN, Tudor MJ, *et al.* A micro electromagnetic generator for vibration energy harvesting. *J Micromech Microeng* 2007;17:1257–65. https://doi.org/10.1088/0960-1317/17/7/007.

[257] Rome LC, Flynn L, Goldman EM, Yoo TD. Generating electricity while walking with loads. *Am Assoc Adv Sci* 2005; 309. https://doi.org/10.1126/science.1111063.

[258] He X, Wen Q, Sun Y, Wen Z. A low-frequency piezoelectric-electromagnetic-triboelectric hybrid broadband vibration energy harvester. *Nano Energy* 2017;40:300–7. https://doi.org/10.1016/j.nanoen.2017.08.024.

[259] Khaligh A, Zeng P, Zheng C. Kinetic energy harvesting using piezoelectric and electromagnetic technologies: State of the art. *IEEE Trans Ind Electron* 2010;57:850–60. https://doi.org/10.1109/TIE.2009.2024652.

[260] Zhang Y, Cai CS, Kong B. A low frequency nonlinear energy harvester with large bandwidth utilizing magnet levitation. *Smart Mater Struct* 2015;24:45019. https://doi.org/10.1088/0964-1726/24/4/045019.

[261] Wang L, Yuan FG. Vibration energy harvesting by magnetostrictive material. *Smart Mater Struct* 2008;17:045009. https://doi.org/10.1088/0964-1726/17/4/045009.

[262] Wang L, Chen S, Zhou W, Xu TB, Zuo L. Piezoelectric vibration energy harvester with two-stage force amplification. *J Intell Mater Syst Struct* 2017;28:1175–87. https://doi.org/10.1177/1045389X16667551.

[263] Qian F, Xu TB, Zuo L. Piezoelectric energy harvesting from human walking using a two-stage amplification mechanism. *Energy* 2019;189:116140. https://doi.org/10.1016/j.energy.2019.116140.

[264] Edwards B, Hu PA, Aw KC. Validation of a hybrid electromagnetic–piezoelectric vibration energy harvester. *Smart Mater Struct* 2016;25:0. https://doi.org/10.1088/0964-1726/25/5/055019.

[265] Toyabur RM, Kim JW, Park JY. A hybrid piezoelectric and electromagnetic energy harvester for scavenging low frequency ambient vibrations. *J Phys Conf Ser* 2018;1052:0–4. https://doi.org/10.1088/1742-6596/1052/1/012051.

[266] Li Z, Li T, Yang Z, Naguib HE. Toward a 0.33 W piezoelectric and electromagnetic hybrid energy harvester: Design, experimental studies and self-powered applications. *Appl Energy* 2019;255:113805. https://doi.org/10.1016/j.apenergy.2019.113805.

[267] Hamid R, Yuce MR. A wearable energy harvester unit using piezoelectric–electromagnetic hybrid technique. *Sensors Actuators A Phys* 2017;257:198–207. https://doi.org/10.1016/j.sna.2017.02.026.

[268] Mann BP, Sims ND. Energy harvesting from the nonlinear oscillations of magnetic levitation. *J Sound Vib* 2009;319:515–30. https://doi.org/10.1016/j.jsv.2008.06.011.

[269] Masoumi M, Wang Y. Repulsive magnetic levitation-based ocean wave energy harvester with variable resonance: Modeling, simulation and experiment. *J Sound Vib* 2016;381:192–205. https://doi.org/10.1016/j.jsv.2016.06.024.

[270] Mitcheson PD, Green TC, Yeatman EM, Holmes AS. Architectures for vibration-driven micropower generators. *J Microelectromech Syst* 2004;13:429–40. https://doi.org/10.1109/JMEMS.2004.830151.

[271] He J, Fu Z-F. Basic vibration theory. *Modal Anal* 2001; 49–78. https://doi.org/10.1016/b978-075065079-3/50003-6.

[272] Sterken T, Baert K, Van Hoof C, Puers R, Borghs G, Fiorini P. Comparative modelling for vibration scavengers. *Proc IEEE Sens* 2004;3:1249–52. https://doi.org/10.1109/icsens.2004.1426407.

[273] Khan SFU. *Vibration-based Electromagnetic Energy Harvesters for MEMS Applications*. Vancouver: The University of British Columbia; 2011. https://doi.org/10.14288/1.0071751.

[274] Kaźmierski TJ, Beeby S. *Energy Harvesting Systems: Principles, Modeling And Applications*. New York, NY: Springer; 2011. https://doi.org/10.1007/978-1-4419-7566-9.

[275] Elements of design of magnetic separation equipment. *Magnetic Techniques for the Treatment of Materials*. Johannesburg, South Africa: Dordrecht: Springer; 2006, pp. 251–318. https://doi.org/10.1007/1-4020-2107-0_4.

[276] Zhu M, Worthington E, Njuguna J. Analyses of power output of piezoelectric energy harvesting devices directly connected to a resistive load using a coupled piezoelectric-circuit finite element method. *IEEE Trans Ultrason Ferroelectr Freq Control* 2009;56:1309–17.

[277] Jiang XY, Zou HX, Zhang WM. Design and analysis of a multi-step piezoelectric energy harvester using buckled beam driven by magnetic excitation. *Energy Convers Manag* 2017;145:129–37. https://doi.org/10.1016/j.enconman.2017.04.088.

[278] Lee H, Choi TK, Lee YB, *et al.* A graphene-based electrochemical device with thermoresponsive microneedles for diabetes monitoring and therapy. *Nat Nanotechnol* 2016;11:566–72. https://doi.org/10.1038/nnano.2016.38.

[279] Bai P, Zhu G, Jing Q, *et al.* Membrane-based self-powered triboelectric sensors for pressure change detection and its uses in security surveillance and healthcare monitoring. *Adv Funct Mater* 2014;24:5807–13. https://doi.org/10.1002/adfm.201401267.

[280] Tian Z, He J, Chen X, *et al.* Core-shell coaxially structured triboelectric nanogenerator for energy harvesting and motion sensing. *RSC Adv* 2018;8:2950–7. https://doi.org/10.1039/c7ra12739a.

[281] Larcher D, Tarascon JM. Towards greener and more sustainable batteries for electrical energy storage. *Nat Chem* 2015;7:19–29. https://doi.org/10.1038/nchem.2085.

[282] Lisbona D, Snee T. A review of hazards associated with primary lithium and lithium-ion batteries. *Process Saf Environ Prot* 2011;89:434–42. https://doi.org/10.1016/j.psep.2011.06.022.

[283] Recham N, Chotard JN, Dupont L, *et al.* A 3.6 V lithium-based fluorosulphate insertion positive electrode for lithium-ion batteries. *Nat Mater* 2010;9:68–74. https://doi.org/10.1038/nmat2590.

[284] Armand M, Tarascon JM Building better batteries. *Nature* 2008;451:652–57.

[285] Han S, Kim J, Won SM, *et al.* Battery-free, wireless sensors for full-body pressure and temperature mapping. *Sci Transl Med* 2018;10. https://doi.org/10.1126/scitranslmed.aan4950.

[286] Shi B, Liu Z, Zheng Q, *et al.* Body-integrated self-powered system for wearable and implantable applications. *ACS Nano* 2019;13:6017–24. https://doi.org/10.1021/acsnano.9b02233.

[287] Guo H, Chen J, Yeh MH, *et al.* An ultrarobust high-performance triboelectric nanogenerator based on charge replenishment. *ACS Nano* 2015;9:5577–84. https://doi.org/10.1021/acsnano.5b01830.

[288] Khan FU, Ahmad S. Flow type electromagnetic based energy harvester for pipeline health monitoring system. *Energy Convers Manag* 2019; 200:112089. https://doi.org/10.1016/j.enconman.2019.112089.

[289] Zheng L, Cheng G, Chen J, *et al.* A hybridized power panel to simultaneously generate electricity from sunlight, raindrops, and wind around the clock. *Adv Energy Mater* 2015;5:1–8. https://doi.org/10.1002/aenm.2015 01152.

[290] Kymissis J, Kendall C, Paradiso J, Gershenfeld N. Parasitic power harvesting in shoes. *Int Symp Wearable Comput Dig Pap 1998*;1998 October: pp. 132–9. https://doi.org/10.1109/ISWC.1998.729539.

[291] Park JH, Wu C, Sung S, Kim TW. Ingenious use of natural triboelectrification on the human body for versatile applications in walking energy harvesting and body action monitoring. *Nano Energy* 2019;57:872–8. https://doi.org/10.1016/j.nanoen.2019.01.001.

[292] Lukowicz P, Anliker U, Ward J, Troster G, Hirt E, Neufelt C. AMON: A wearable medical computer for high risk patients. *Proc – Int Symp Wearable Comput ISWC*, vol. 2002 – January 2002, pp. 133–4. https://doi.org/10.1109/ISWC.2002.1167230.

[293] Nagae D, Mase A. Measurement of heart rate variability and stress evaluation by using microwave reflectometric vital signal sensing. *Rev Sci Instrum* 2010;81. https://doi.org/10.1063/1.3478017.

[294] Steele BG, Belza B, Cain K, Warms C, Coppersmith J, Howard JE. Bodies in motion: Monitoring daily activity and exercise with motion sensors in people with chronic pulmonary disease. *J Rehabil Res Dev* 2003;40:45–58. https://doi.org/10.1682/JRRD.2003.10.0045.

[295] Cook DJ, Thompson JE, Prinsen SK, Dearani JA, Deschamps C. Functional recovery in the elderly after major surgery: Assessment of mobility recovery using wireless technology. *Ann Thorac Surg* 2013;96:1057–61. https://doi.org/10.1016/j.athoracsur.2013.05.092.

[296] Li H, Tian C, Deng ZD. Energy harvesting from low frequency applications using piezoelectric materials. *Appl Phys Rev* 2014;1:0–20. https://doi.org/10.1063/1.4900845.

[297] Fan K, Yu B, Zhu Y, Liu Z, Wang L. Scavenging energy from the motion of human lower limbs via a piezoelectric energy harvester. *Int J Mod Phys B* 2017;31:1741011. https://doi.org/10.1142/S0217979217410119.

[298] Moro L, Benasciutti D. Harvested power and sensitivity analysis of vibrating shoe-mounted piezoelectric cantilevers. *Smart Mater Struct* 2010; 19. https://doi.org/10.1088/0964-1726/19/11/115011.

[299] Fan K, Liu Z, Liu H, Wang L, Zhu Y, Yu B. Scavenging energy from human walking through a shoe-mounted piezoelectric harvester. *Appl Phys Lett* 2017;110. https://doi.org/10.1063/1.4979832.

[300] Li K, He Q, Wang J, Zhou Z, Li X. Wearable energy harvesters generating electricity from low-frequency human limb movement. *Microsyst Nanoeng* 2018;4. https://doi.org/10.1038/s41378-018-0024-3.

[301] Lu B, Chen Y, Ou D, *et al.* Ultra-flexible piezoelectric devices integrated with heart to harvest the biomechanical energy. *Sci Rep* 2015;5:1–9. https://doi.org/10.1038/srep16065.

[302] Wang ZL. Triboelectric nanogenerators as new energy technology for self-powered systems and as active mechanical and chemical sensors. *ACS Nano* 2013;7:9533–57. https://doi.org/10.1021/nn404614z.

[303] Mahmud A, Khan AA, Islam S, Voss P, Ban D. Integration of organic/inorganic nanostructured materials in a hybrid nanogenerator enables efficacious energy harvesting via mutual performance enhancement. *Nano Energy* 2019;58:112–20. https://doi.org/10.1016/j.nanoen.2019.01.023.

[304] Wang X, Niu S, Yi F, *et al.* Harvesting ambient vibration energy over a wide frequency range for self-powered electronics. *ACS Nano* 2017;11:1728–35. https://doi.org/10.1021/acsnano.6b07633.

[305] Chen J, Zhu G, Yang W, *et al.* Harmonic-resonator-based triboelectric nanogenerator as a sustainable power source and a self-powered active vibration sensor. *Adv Mater* 2013;25:6094–9. https://doi.org/10.1002/adma.201302397.

[306] Dong K, Peng X, Wang ZL. Fiber/fabric-based piezoelectric and triboelectric nanogenerators for flexible/stretchable and wearable electronics and artificial intelligence. *Adv Mater* 2019;1902549:1902549. https://doi.org/10.1002/adma.201902549.

[307] García Núñez C, Manjakkal L, Dahiya R. Energy autonomous electronic skin. *npj Flex Electron* 2019;3. https://doi.org/10.1038/s41528-018-0045-x.

[308] Bhatia D, Hwang HJ, Huynh ND, *et al.* Continuous scavenging of broadband vibrations via omnipotent tandem triboelectric nanogenerators with cascade impact structure. *Sci Rep* 2019;9:8223. https://doi.org/10.1038/s41598-019-44683-5.

[309] Liu L, Tang W, Deng C, *et al.* Self-powered versatile shoes based on hybrid nanogenerators. *Nano Res* 2018;11:3972–8. https://doi.org/10.1007/s12274-018-1978-z.

[310] Hou TC, Yang Y, Zhang H, Chen J, Chen LJ, Lin Wang Z. Triboelectric nanogenerator built inside shoe insole for harvesting walking energy. *Nano Energy* 2013;2:856–62. https://doi.org/10.1016/j.nanoen.2013.03.001.

[311] Liu H, Zhong J, Lee C, Lee SW, Lin L. A comprehensive review on piezoelectric energy harvesting technology: Materials, mechanisms, and applications. *Appl Phys Rev* 2018;5. https://doi.org/10.1063/1.5074184.

[312] Cheskin M. Shoe with electrostatic and endogenous current conducting insert. US 2019/0183205 A1, 2019.

[313] Li Y, Chen Z, Zheng G, *et al.* A magnetized microneedle-array based flexible triboelectric-electromagnetic hybrid generator for human motion monitoring. *Nano Energy* 2020;69:104415. https://doi.org/10.1016/j.nanoen.2019.104415.

[314] Farhangdoust S, Georgeson G, Ihn JB., Chang FK, Kirigami auxetic structure for high efficiency power harvesting in self-powered and wireless structural health monitoring systems. *Smart Mater Struct* 2020;30. doi: 10.1088/1361-665X/abcaaf.

[315] Paul K, Amann A, Roy S. Tapered nonlinear vibration energy harvester for powering Internet of Things. *Appl Energy* 2021;283. doi: 10.1016/j.apenergy.2020.116267.

[316] Zhang T, Tang M, Li H, Li J., A multidirectional pendulum kinetic energy harvester based on homopolar repulsion for low-power sensors in new energy driverless buses. *Int J Prec Eng Manuf – Green Technol* 2022;9:603–18. doi: 10.1007/s40684-021-00344-5.

[317] Dahat N, Maskar P, Yadav A Design and development of regenerative shock absorber. *Mater Today Proc* 2023. doi: 10.1016/j.matpr.2023.02.433.

[318] Lafarge B, Grondel S, Delebarre C, Curea O, Richard C. Linear electromagnetic energy harvester system embedded on a vehicle suspension: From modeling to performance analysis. *Energy* 2021;225. doi: 10.1016/j.energy.2021.119991.

[319] Lou H, Wang T, Zhu S. Design, modeling and experiments of a novel biaxial-pendulum vibration energy harvester. *Energy* 2022; 254. doi: 10.1016/j.energy.2022.124431.

[320] Narolia T, Gupta VK, Parinov IA. Design and analysis of a shear mode piezoelectric energy harvester for rotational motion system. *J Adv Dielectr* 2020;10. doi: 10.1142/S2010135X20500083.

[321] Ju S, Ji CH. Impact-based piezoelectric vibration energy harvester. *Appl Energy* 2018;214:139–51. doi: 10.1016/j.apenergy.2018.01.076.

[322] Shi G, Zeng W, Xia Y, *et al.*, A floating piezoelectric electromagnetic hybrid wave vibration energy harvester actuated by a rotating wobble ball. *Energy* 2023;270. doi: 10.1016/j.energy.2023.126808.

[323] Rosso M, Corigliano A, Ardito R, Numerical and experimental evaluation of the magnetic interaction for frequency up-conversion in piezoelectric vibration energy harvesters. *Meccanica* 2022;57:1139–54. doi: 10.1007/s11012-022-01481-0.

[324] Zeng P, Khaligh A, A permanent-magnet linear motion driven kinetic energy harvester. *IEEE Trans Industr Electr.*, 2013; 60:5737–46. doi: 10.1109/TIE.2012.2229674.

[325] Halim MA, Rantz R, Zhang Q, Gu L, Yang K, Roundy S. An electromagnetic rotational energy harvester using sprung eccentric rotor, driven by pseudo-walking motion. *Appl Energy* 2018;217:66–74. doi: 10.1016/j.apenergy.2018.02.093.

[326] Ham SS, Lee G-J, Hyeon DY, Kim Y-g. Kinetic motion sensors based on flexible and lead-free hybrid piezoelectric composite energy harvesters

with nanowires-embedded electrodes for detecting articular movements. *Compos B Eng* 2021;212. doi: 10.1016/j.compositesb.2021.108705.

[327] Sezer N, Koç M. A comprehensive review on the state-of-the-art of piezoelectric energy harvesting. *Nano Energy* 2021; 80:105567. doi: 10.1016/j.nanoen.2020.105567.

[328] Zhang R, Olin H. Material choices for triboelectric nanogenerators: A critical review. *EcoMat* 2020; 2. doi: 10.1002/eom2.12062.

[329] Wang H, Shan X, Xie T. Performance optimization for cantilevered piezoelectric energy harvester with a resistive circuit. *2012 IEEE Int Conf Mechatr Autom*, Piscataway, NJ: IEEE; 2012, pp. 2175–80. doi: 10.1109/ICMA.2012.6285680.

[330] Zhang X, Guo Y, Zhu F, Chen X, Tian H, Xu H. A linear-arc composite beam piezoelectric energy harvester modeling and finite element analysis. *Micromachines (Basel)* 2022;13. doi:10.3390/mi13060848.

[331] Xie Z, Liu L, Huang W, *et al.*, A multimodal E-shaped piezoelectric energy harvester with a built-in bistability and internal resonance. *Energy Convers Manag* 2023;278. doi: 10.1016/j.enconman.2023.116717.

[332] Upadrashta D, Yang Y. Finite element modeling of nonlinear piezoelectric energy harvesters with magnetic interaction. *Smart Mater Struct* 2015;24. doi: 10.1088/0964-1726/24/4/045042.

[333] Gao S, Gain AK, Zhang L. A metamaterial for wearable piezoelectric energy harvester. *Smart Mater Struct* 2020;30. doi: 10.1088/1361-665X/abca09.

[334] Du C, Liu P, Jin C, *et al.*, Evaluation of the piezoelectric and mechanical behaviors of asphalt pavements embedded with a piezoelectric energy harvester based on multiscale finite element simulations. *Constr Build Mater* 2022;333. doi: 10.1016/j.conbuildmat.2022.127438.

[335] Narolia T, Gupta VK, Parinov IA. Design and experimental study of rotary-type energy harvester. *J Intell Mater Syst Struct* 2020;31:1594–603. doi: 10.1177/1045389X20930085.

[336] Fan S, Tang Y, Liu H, *et al.*, Design optimization of microfabricated coils for volume-limited miniaturized broadband electromagnetic vibration energy harvester. *Energy Convers Manag* 2022;271. doi: 10.1016/j.enconman.2022.116299.

[337] Gupta RK, Shi Q, Dhakar L, Wang T, Heng CH, Lee C. Broadband energy harvester using non-linear polymer spring and electromagnetic/triboelectric hybrid mechanism. *Sci Rep* 2017;7. doi: 10.1038/srep41396.

[338] Augustyniak M, Usarek Z. Finite element method applied in electromagnetic NDTE: A review. *J Nondestr Eval* 2016;35:39. doi: 10.1007/s10921-016-0356-6.

[339] Magomedov IA, Sebaeva ZS, Comparative study of finite element analysis software packages. *J Phys Conf Ser* 2020;1515:032073. doi: 10.1088/1742-6596/1515/3/032073.

[340] David Müzel S, Bonhin EP, Guimarães NM, Guidi ES. Application of the finite element method in the analysis of composite materials: A review. *Polymers (Basel)* 2020;12:818. doi: 10.3390/polym12040818.

[341] Liu M, Qian F, Mi J, Zuo L. Biomechanical energy harvesting for wearable and mobile devices: State-of-the-art and future directions. *Appl Energy* 2022;321. https://doi.org/10.1016/j.apenergy.2022.119379.

[342] Maiti S, Karan SK, Kim JK, Khatua BB. Nature driven bio-piezoelectric/triboelectric nanogenerator as next-generation green energy harvester for smart and pollution free society. *Adv Energy Mater* 2019;9:1803027.

[343] Dagdeviren C, Joe P, Tuzman OL, *et al.* Recent progress in flexible and stretchable piezoelectric devices for mechanical energy harvesting, sensing and actuation. *Extrem Mech Lett* 2016;9:269–81.

[344] Bagherzadeh R, Abrishami S, Shirali A, Rajabzadeh AR. Wearable and flexible electrodes in nanogenerators for energy harvesting, tactile sensors, and electronic textiles: Novel materials, recent advances, and future perspectives. *Mater Today Sustain* 2022:100233.

[345] Guan X, Xu B, Wu M, Jing T, Yang Y, Gao Y. Breathable, washable and wearable woven-structured triboelectric nanogenerators utilizing electrospun nanofibers for biomechanical energy harvesting and self-powered sensing. *Nano Energy* 2021;80:105549.

[346] Iqbal M, Nauman MM, Khan FU, *et al.* Vibration-based piezoelectric, electromagnetic, and hybrid energy harvesters for microsystems applications: A contributed review. *Int J Energy Res* 2021;45:65–102. https://doi.org/10.1002/er.5643.

[347] Zhou X, Liu G, Han B, Liu X. Different kinds of energy harvesters from human activities. *Int J Energy Res* 2021;45:4841–70.

[348] Uchida TK, Delp SL. *Biomechanics of Movement: The Science of Sports, Robotics, and Rehabilitation*. Cambridge, MA: MIT Press; 2021.

[349] Zou Y, Bo L, Li Z. Recent progress in human body energy harvesting for smart bioelectronic system. *Fundam Res* 2021;1:364–82.

[350] Shahab H, Abbas T, Sardar MU, Basit A, Waqas MM, Raza H. Internet of Things implications for the adequate development of the smart agricultural farming concepts. *Big Data Agric* 2020;3:12–7.

[351] Gupta A, Chaurasiya VK. Reinforcement learning based energy management in wireless body area network: A survey. 2019 *IEEE Conf Inf Commun Technol*, Piscataway, NJ: IEEE; 2019, pp. 1–6.

[352] Saidjon K, Bakhrom U. Energy-saving materials in residential architecture. *Am J Eng Technol* 2021;3:44–7.

[353] Qaisar F, Shahab H, Iqbal M, Sargana HM, Aqeel M, Qayyum MA. Recent trends in cloud computing and IoT platforms for IT management and development: A review. *Pak J Eng Technol* 2023;6:98–105.

[354] Shuvo MMH, Titirsha T, Amin N, Islam SK. Energy harvesting in implantable and wearable medical devices for enduring precision healthcare. *Energies* 2022;15:7495.

[355] Khan AS, Khan FU. A survey of wearable energy harvesting systems. *Int J Energy Res* 2022;46:2277–329.

[356] Ramalingam L, Mariappan S, Parameswaran P, *et al.* The advancement of radio frequency energy harvesters (RFEHs) as a revolutionary approach for

solving energy crisis in wireless communication devices: A review. *IEEE Access* 2021;9:106107–39.

[357] Jurado UT, Pu SH, White NM. Wave impact energy harvesting through water-dielectric triboelectrification with single-electrode triboelectric nanogenerators for battery-less systems. *Nano Energy* 2020;78:105204.

[358] Mishu MK, Rokonuzzaman M, Pasupuleti J, *et al.* Prospective efficient ambient energy harvesting sources for IoT-equipped sensor applications. *Electronics* 2020;9:1345.

[359] Roy S, Azad ANMW, Baidya S, Alam MK, Khan FH. Powering solutions for biomedical sensors and implants inside human body: A comprehensive review on energy harvesting units, energy storage, and wireless power transfer techniques. *IEEE Trans Power Electron* 2022.

[360] Zhou Z, Weng L, Tat T, *et al.* Smart insole for robust wearable biomechanical energy harvesting in harsh environments. *ACS Nano* 2020;14:14126–33. https://doi.org/10.1021/acsnano.0c06949.

[361] Gupta S, Kumar M, Singh G, Chanda A. Development of a novel footwear based power harvesting system. *e-Prime – Adv Electr Eng Electron Energy* 2023:100115.

[362] Kashfi M, Fakhri P, Amini B, *et al.* A novel approach to determining piezoelectric properties of nanogenerators based on PVDF nanofibers using iterative finite element simulation for walking energy harvesting. *J Ind Text* 2022;51:531S–53S.

[363] Shi H, Liu Z, Mei X. Overview of human walking induced energy harvesting technologies and its possibility for walking robotics. *Energies* 2020;13:22. https://doi.org/10.3390/en13010086.

[364] Megdich A, Habibi M, Laperrière L. A review on 3D printed piezoelectric energy harvesters: Materials, 3D printing techniques, and applications. *Mater Today Commun* 2023:105541.

[365] Rashid ARA, Sarif NIM, Ismail K. Development of smart shoes using piezoelectric material. *Malaysian J Sci Heal Technol* 2021.

[366] Shah N, Kamdar L, Gokalgandhi D, Mehendale N. Walking pattern analysis using deep learning for energy harvesting smart shoes with IoT. *Neural Comput Appl* 2021;33:11617–25.

[367] Cai M, Liao W-H. Design, modeling, and experiments of electromagnetic energy harvester embedded in smart watch and wristband as power source. *IEEE/ASME Trans Mechatron* 2020;26:2104–14.

[368] Maharjan P, Cho H, Rasel MS, Salauddin M, Park JY. A fully enclosed, 3D printed, hybridized nanogenerator with flexible flux concentrator for harvesting diverse human biomechanical energy. *Nano Energy* 2018;53:213–24. https://doi.org/10.1016/j.nanoen.2018.08.034.

[369] Fernández-Caramés TM, Fraga-Lamas P. Towards the Internet of smart clothing: A review on IoT wearables and garments for creating intelligent connected e-textiles. *Electronics* 2018;7:405.

[370] Júnior HLO, Neves RM, Monticeli FM, Dall Agnol L. Smart fabric textiles: Recent advances and challenges. *Textiles* 2022;2:582–605.

[371] Du K, Lin R, Yin L, Ho JS, Wang J, Lim CT. Electronic textiles for energy, sensing, and communication. *iScience* 2022:104174.

[372] Liu Y, Pharr M, Salvatore GA. Lab-on-skin: A review of flexible and stretchable electronics for wearable health monitoring. *ACS Nano* 2017;11:9614–35.

[373] Lu K, Yang L, Seoane F, Abtahi F, Forsman M, Lindecrantz K. Fusion of heart rate, respiration and motion measurements from a wearable sensor system to enhance energy expenditure estimation. *Sensors* 2018;18:3092.

[374] Larson W, Wertz J, *Spacecraft Mission Analysis and Design, Space Technology Library*, Wiley Series, Dordrecht: Springer; 2002. ISSN 0924-4263.

[375] Cooper SM. Control of a satellite based photovoltaic array for optimum power draw, WPI thesis, Worcester Polytechnic Institute, Worcester, Massachusetts, USA, April 2008.

[376] Jaziri N, Boughamoura A, Müller J, Mezghani B, Tounsi F, Ismail M. A comprehensive review of thermoelectric generators: Technologies and common applications, *Energy Rep* 2020; 6:264–87.

[377] Tritt TM. Thermoelectric materials: Principles, structure, properties, and applications. *Encycl Mater: Sci Technol* 2002, 1–11.

[378] (Online resource) 500 Watt solar AMTEC power system for small spacecraft.

[379] Johnson G, Hunt ME, Determan WR, HoSang PA, Ivanenok J, Schuller M. Design and integration of a solar AMTEC power system with an advanced global positioning satellite. *IECEC 96. Proc 31st Intersociety Energy Convers Eng Conf*, vol. 1, Washington, DC, USA, 1996, pp. 623–28, doi: 10.1109/IECEC.1996.552965

[380] Operating manual of "8550 Thermoelectric Generator Operating Manual, Global Thermoelectric". Available online at: Global thermoelectric 8550 Manuals | ManualsLib.

[381] Bilen S, McTernan J, Gilchrist B, Bell I, Voronka N, Hoyt R Electrodynamic tethers for energy harvesting and propulsion on space platforms. *AIAA-2010-8844, AIAA SPACE 2010 Conf Exposition*, Anaheim, California, August 30–2, 2010.

[382] McTernan J, Gilchrist B, Bilen S, Hoyt R, Voronka N, Bell I. Development of a modeling capability for energy harvesting modules in electrodynamic tether systems. *AIAA SPACE 2011 Conf Exposition*, Long Beach, California, September 27–29, 2011, AIAA-2011-7323.

[383] Smith B, Sanmartin JR, Martinez-Sanchez M, Ahedo E. Bare wire anodes for electrodynamic tethers. *J Propul Pow* 1993; 9: 353–60.

[384] Bilén SG, McTernan JK, Gilchrist BE, Bell IC, Voronka NR, Hoyt RP. Energy harvesting electrodynamic tethers, *11th Spacecraft Charging Technology Conference*, Albuquerque, New Mexico, 20–24 September 2010.

[385] Bilen SG, McTernan JK, Gilchrist BE, *et al.* Harnessing the "orbital battery" for propulsion via energy harvesting electrodynamic, *IEPC-2011-226*, 2011.

[386] Bilen SG, McTernan JK, Gilchrist BE, *et al.* The potential of miniature electrodynamic tethers to enhance capabilities of femtosatellites, *IEPC-2011-054*, 2011.

[387] Win KK, Dasgupta S, Panda SK. An optimized MPPT circuit for thermoelectric energy harvester for low power applications. *8th Int Conf Power Electron – ECCE Asia*, Jeju, Korea (South), 2011, pp. 1579–84, doi: 10.1109/ICPE.2011.5944535.

[388] Collignon JM, Quement A, Picdi BB, Rmili B, Espinosa A. RFID and RF harvesting wireless sensor network platform for launcher application. *2015 IEEE Int Conf Wirel Space Extreme Environ (WiSEE)*, Orlando, FL, 2015, pp. 1–6, doi: 10.1109/WiSEE.2015.7393101

[389] Zakaria Z, Zainuddin NA, Husain MN, Abd Aziz MZA, Mutalib MA, Othman AR. Current developments of RF energy harvesting system for wireless sensor networks. *Adv Inform Sci Serv Sci (AISS)* 2013;5. doi:10.4156/ AISS.vol5.issue11.39

[390] Collignon J-M, Quement A, Baron B, Rmili B. RF energy harvester optimized for wireless sensor network in launcher application. *2017 IEEE Int Conf Wirel Space Extreme Environ (WiSEE)*, Montreal, QC, Canada, 2017, pp. 48–51, doi: 10.1109/WiSEE.2017.8124891.

[391] Stoopman M, Keyrouz S, Visser HJ, Philips K, Serdijn WA. Co-design of a CMOS rectifier and small loop antenna for highly sensitive RF energy harvesters. *IEEE J Solid-State Circuits* 2014;49:622–34.

[392] Le T., Mayaram K, Fiez T. Efficient far-field radio frequency energy harvesting for passively powered sensor networks. *IEEE J Solid-State Circuits* 2008;43:1287–1302. doi:10.1109/JSSC.2008.920318.

[393] Visser HJ, Pop V, Op het Veld, JHG, Vullers RJM. Remote RF battery charging. *PowerMEMS 2010: 10th Int Workshop Micro Nanotechnol Power Gener Energy Convers Appl*, Leuven, Belgium, 30 November–3 December 2010, 2011 April, London: IET.

[394] Devi K, Din N, Chakrabarty C. Optimization of the voltage doubler stages in an RF-DC convertor module for energy harvesting. *Circ Syst* 2012;3:216–22. doi: 10.4236/cs.2012.33030.

[395] Md Din N., Chakrabarty CK, Bin Ismail A, Devi KKA, Chen WY. Design of RF energy harvesting system for energizing low power devices. *Prog Electromag Res* 2012;132:49–69. doi:10.2528/PIER12072002.

# Index

Printed in the USA
CPSIA information can be obtained
at www.ICGtesting.com
JSHW011311240624
65298JS00003B/207

9 781839 534973